ID0459659

THE MAN WITH THE BIONIC BRAIN
and OTHER VICTORIES OVER PARALYSIS

The Man with *the* Bionic Brain

and Other Victories over Paralysis

Jon Mukand, MD, PhD

ISBN 978-1-61374-055-2

Interior design: Sarah Olson

Library of Congress Cataloging-in-Publication Data
Mukand, Jon, 1959–
  The Man with the bionic brain : and other victories over paralysis / Jon Mukand. —
1st ed.
      p. cm.
  Summary: "After he was stabbed, Matthew Nagle, a former high school football star,
made scientific history when neurosurgeons implanted a microelectrode in his brain.
Using BrainGate technology, Matt could merely think about moving a computer
cursor—and it moved. He controlled the lights, manipulated his prosthetic hand,
turned the TV off and on, and played video games, all just by thinking. In The Man
with the Bionic Brain, Dr. Jon Mukand, Matt's research physician and a specialist in
rehabilitation medicine, weaves together the stories of Matt and other survivors of
stroke, spinal injuries, and brain trauma; his relationship with them; and the technology
that is working miracles. Advances in biomedicine are a matter of life and death for
the patients, but they are often caught in the crossfire of cultural wars over the limits
of science, from animal studies to the FDA, financing, and publication. In an era of
wounded veterans and an aging population, The Man with the Bionic Brain provides
inspiration and insight into the possibilities of technology and explores cutting-edge
human research and the attendant ethical, political, social, and financial controversies.
Ultimately, the book is about people with disabilities realizing their dreams of healing
their damaged bodies and regaining any measure of control"— Provided by publisher.
  ISBN 978-1-61374-055-2 (hardback)
  1. Nagle, Matthew, 1979-2007—Health. 2. Quadriplegia—Treatment—Technological
innovations. 3. Paralytics—Rehabilitation. 4. Brain-computer interfaces. 5. Implants,
Artificial. 6. Neurosciences. I. Title.

RC406.Q33M85 2012
617.4'82044—dc23

2012008154

For Giselle, Nita, and Jacob,

who keep my neurons firing

&

In loving memory of my parents,

Ivy and Eric Mukand

Contents

Prologue

⸺⸺

It is Sunday afternoon, and my wheelchaired patients at the Southern New England Rehabilitation Center are sitting in a circle, enjoying a session of therapeutic recreation. All of them are paralyzed to some extent, disabled, unable to take care of their basic daily needs. They have worked hard during the week to recover from the effects of strokes, hip fractures, brain injuries, amputations, spinal cord injuries (SCI), and other conditions. They have spent hours and hours relearning activities that most people take for granted, like bathing, dressing, and walking. With the help of their rehabilitation therapists and nurses, they are making the most of whatever movement is left in their bodies.

Anna Iacono, an elderly woman whose right side is paralyzed due to a stroke, tries to walk again. Early in her stay, she struggled just to move around in bed. Now she wears a plastic brace for her foot and uses a four-legged walker while her physical therapist retrains her steps.

Next to Anna in the circle is a middle-aged man who fell from a roof and crushed his spinal cord. His legs are paralyzed, but he uses his muscular arms to move from his bed to his wheelchair and propel himself. He is still learning how to manage his bowel and bladder functions, which are also affected by spinal cord injuries.

My patients are listening to the disco song "YMCA," by the Village People, while a recreation therapist leads them in a dance. The infectious rhythm of this party favorite encourages even wallflowers to get on the dance floor and swing their arms around, swaying with the raucous crowd. I imagine that the therapist chose this song because it gets people to dance with their arms and legs, and everyone here can move at least one limb. Anna can use her left arm and leg, while her fellow patient is able to use his arms. Anna holds her left arm aloft to form the Y as she struggles to raise her right upper limb. The man with paraplegia reaches over to raise her arm, in the process forming a Y for both of them. He has no trouble bending his elbows so that his hands meet in front to form the M, but Anna can only make an effort. For the C, Anna relies on her left arm. Her partner forms the A with his hands above his head; but he cannot help Anna, so she raises just her left arm.

Now the therapist encourages them to sing along with the letters. Anna's voice is hoarse, and some words are broken because of her stroke. But she does her best to belt out the song, refusing to be left out of the fun. Even though the dance reminds my patients of their paralysis, they enjoy this therapy, knowing that no one will judge their awkward movements.

"Paralysis," a word derived from the Greek, refers to the loss of movement. It is typically caused by damage to upper motor neurons in the central nervous system (the brain and spinal cord) or lower motor neurons—peripheral nerves—that originate in the spinal cord and course through the limbs to the fingers and toes. Injuries may occur at any point in the chain of nerves originating in the brain and ending at the muscles. Strokes and spinal cord injuries result from damage to upper motor neurons, which travel down from the brain inside the spinal cord, to connect with and control the lower motor neurons, which move the limbs. Whether nerves are affected in the brain or spinal cord or limbs, the result is paralysis. Sometimes I try to imagine what life is like for my patients. For someone with complete paralysis due to a spinal cord injury at the neck, attempting to move is probably like my trying to move a wall. After a person suffers a stroke that paralyzes half the body, trying to walk is like pushing a table across a room.

In the United States, select rehabilitation centers receive funding from the National Institute on Disability and Rehabilitation Research. Part of the reason I specialized in rehabilitation medicine was that I attended the Medical College of Wisconsin, which housed such a center. There I met a young woman with a spinal cord injury whose sad brown eyes haunt me to this day. We were about the same age, and I sympathized with her physical and psychological distress. Another memorable patient had a severe right-brain stroke, which limited her awareness of her left side—and her recovery.

To better understand people with catastrophic disabilities, I completed a research project on their perception of time. My results showed that they were cut off from the past, uncertain about the future, and adrift in the present. But as a medical student in 1984, I felt there were many reasons to hope. I admired the patients and their devoted therapists, nurses, and physicians. Specialists who designed and fabricated braces, artificial limbs, and other devices were taking advantage of technological breakthroughs. And physician scientists were optimistic because of research in neural plasticity, the brain's ability to adapt by changing the structure and function of its nerves. In 1925, Albrecht Bethe, a physician at the University of Frankfurt, was among the first scientists to describe the concept of neural plasticity. Bethe was fascinated with the structure and function of nerves: their microanatomy, interconnections, electrical activity, and regeneration. He studied brain reorganization in crabs with one, two, or three amputated limbs. Remarkably, the crabs were able to move despite the amputations, supporting Bethe's view that neural plasticity is related to brain reorganization and adaptation to new circumstances.[1]

Until my experience at the rehabilitation center, I held a traditional view that brain recovery after strokes and traumatic injuries was limited. Hearing about neural plasticity was an epiphany. I specialized in rehabilitation medicine to help people with disabilities survive their medical problems, overcome their limitations, and thrive for the rest of their lives. Over the next quarter century, I saw remarkable advances that ranged from medications that prevent blood clots to implanted devices that alleviate bladder dysfunction after spinal injuries. Today, as a clinical faculty member at Brown University and Tufts University, I often engage in research to address the medical problems of people with disabilities, such as spasticity, osteoporosis, and blood clots. Much more remains to be done. Neural plasticity remains a fundamental concept and a driving force in rehabilitation medicine.

Anna, her fellow patients, and millions of other people like them are desperate to regain any movement, adapt to their losses, and rebuild their lives. They are willing to spend countless hours with their rehabilitation therapists to make the most of any movement in their damaged bodies. Some are willing to spend thousands of dollars on experimental surgeries and stem cell therapies. The economic costs of spinal cord injuries and strokes are enormous; the human costs are unquantifiable.

This book describes a handful of extraordinary people with spinal injuries and strokes who succeeded, in various ways, in their struggles with paralysis. Their programs of rehabilitation have involved traditional therapies as well as high-tech devices that seem like the products of science fiction.

Most of this text is about Matthew Nagle, a completely paralyzed young man who received a brain implant that was connected to a computer by a fiber-optic cable. I served as a clinical investigator for a research study on this brain-computer interface, the BrainGate. If the experiment succeeded, Matt would merely *think* about controlling the cursor, and he would enter the digital world; if the BrainGate worked, he would control a television or the room temperature or medical robots—with only his thoughts. With tetraplegia (also known as quadriplegia), Matt, the most severely affected of all the patients in this book, is an archetypal figure who represents all people with disabilities. Matt's odyssey extends throughout this book and is chronologically organized—as noted in the chapter subtitles—after his BrainGate surgery.

Interspersed throughout Matt's narrative are the stories of other remarkable patients with strokes and spinal injuries who use traditional, electronic, and robotic devices. Their stories are thematically presented, progressing from the least to the most complex devices. First, Anna Iacono uses a standard plastic brace, Patricia Wines an electronic foot brace, and Linda Holmes a robotic knee brace. The next two narratives discuss upper-extremity devices used after a stroke: Kathy Spencer has an electronic-stimulation splint for her hand, and Garrett Mendez employs a robotic elbow brace. After suffering a spinal cord injury, Floyd Morrow uses robotic braces for both lower extremities. For a similar injury,

Jennifer French has electrodes implanted in her legs, and I describe her in connection with Matt's desire for arm implants. In the epilogue, I briefly discuss the latest victories over paralysis, including an implanted spinal cord stimulator that helped a former baseball player with leg paralysis to walk.

Many other diagnoses—traumatic brain injuries, multiple sclerosis, brain tumors, neuropathies, and Parkinson's disease, to name a few—result in paralysis and disability. The technologies described in this book can be used to help people with these and other paralyzing conditions. But I have focused on people with strokes and spinal cord injuries because they cover a wide array of demographics, causes, severity, disabilities, and societal implications. This spectrum helps me do justice to the diverse experiences of people with disability.

Stroke affects older people (average age seventy)[2] at a time when they have lived productive lives and hope to enjoy their retirement with their families. All the women in this book who have had strokes are grandmothers, and Anna Iacono is a great-grandmother. Spinal cord injury typically affects young adults (average age thirty-two),[3] striking them down as their family and vocational lives are just beginning. Before his injury, Matt had hoped to become a police officer, like his father; afterward, his odds of even becoming a father were poor.

Strokes are usually due to blood clots in arteries, but some also occur because of bleeding in the brain. Almost eight hundred thousand people have a stroke every year (more often than once per minute), and there are about 7 million survivors in the United States.[4] In contrast, paralysis caused by spinal cord injuries is typically due to trauma (motor-vehicle collisions, diving accidents, or gunshots, for example), pressure from a herniated disc or tumor, or reduced blood flow. There are more than ten thousand spinal injuries annually (once every hour) and around one million survivors in the United States.[5]

Residual effects after a stroke range from no weakness to complete paralysis of a limb. (Strokes may also impair speech, swallowing, vision, and cognitive functions, but this book focuses on paralysis.) A spinal cord injury typically causes paralysis of the lower limbs or all extremities; the nerve pathways that go down from the brain to control bowel, bladder, and sexual function are usually damaged; and the sensory nerves that ascend up to the brain are blocked. In contrast to patients with strokes, a far greater proportion of those with spinal cord injuries are limited to life in wheelchairs.

As for the impact on society, about 19 percent of people with strokes require care in a nursing home, with lifetime costs of about $1 million.[6] Spinal cord injuries are far more expensive. On occasion, attorneys who represent people with spinal cord injuries ask me for a life-care plan that includes the costs of clinician visits, equipment, medications, treatment for complications (such as plastic surgery for skin breakdown), and home modifications or nursing-home care. Because people with SCI are often young and medically complex, their lifetime costs can total more than $5 million.

Cultural factors also affect our views of strokes and spinal cord injuries. A stroke is almost accepted as a part of aging (which is a misguided notion). We do our best to help patients with strokes, but a young paralyzed person is considered a more tragic figure. Medical providers, families, and friends often react by doing whatever is humanly possible to help the victim recover. Our compassion, our pity, and our desire to help young people with disabilities are crystallized in our emotional response to the fictional character of Tiny Tim in *A Christmas Carol*—his tragedy is expected to last for decades, burdening him and his family.

By narrating the experiences of people with strokes and spinal cord injuries, I hope to address the following questions: What does it mean to be paralyzed and disabled? How does disability affect us all? How is cutting-edge research conducted? How does this research change our view of disability? How should clinicians discuss and recommend treatments for paralysis? I also present diverse perspectives in a debate related to Matt's desire for stem cells.

Ultimately, this book is about people with disabilities who have lost control over basic bodily functions; they dream about healing their damaged bodies and regaining any measure of control. Through this narrative, I wish to offer them hope.

I am grateful to the patients who have entrusted me with their health care and personal information. I admire their physical and mental toughness as they struggle with paralysis. Medical science has no magic procedures or medications that can fully repair their damaged nerves, but with their canes, walkers, braces, wheelchairs, electric limbs, brain implants, and sheer willpower, they keep moving ahead.

THE MAN WITH THE BIONIC BRAIN
and OTHER VICTORIES OVER PARALYSIS

At the Gateway to the Brain

Matt, June 22, 2004

A drift on anesthesia, Matthew Nagle lay still in the neurosurgical intensive care unit of Rhode Island Hospital, his shaven scalp discolored by a brown antiseptic and covered by a white turban of gauze dressing. Plastic hoses connected the tube in his throat to the ventilator that hissed and puffed and hissed and puffed into his lungs.

While a high school senior in 1998, Matt had galloped across football fields, two hundred pounds of muscled energy charging at opponents; he set a record for unassisted tackles that was featured in the *Boston Herald*. Now twenty-three, he had a severed spinal cord, and his limbs were completely paralyzed. Swaddled in white hospital sheets and floating on a special bed to prevent bedsores, Matt had just had brain surgery as the first human volunteer for the BrainGate Neural Interface System. Sinuous green waves raced across the monitor as his heart, strong as ever, pumped out twenty liters of blood every minute. There had never been a question about Matt's heart while the BrainGate research team planned his surgery.

I touched Matt's upper shoulder, the only place other than his head and neck where he could still feel human contact. My hand appeared dark brown against his pale skin, which seemed bleached by the fluorescent lights of the hospital.

"Matt, how're you feeling?"

Slowly his eyelids opened, a trace of a smile on his face. Then his eyes closed again, the right faster than the left. I felt an infusion of relief; Matt had just winked at me. My grip loosened on the cold rails of his KinAir bed, and I smiled back at my young patient, who drifted away, back into postoperative sleep.

The pill-sized BrainGate implant was now on the surface of Matt's right brain (see figure 1.1 in the gallery). A hundred microelectrodes—each thinner than a hair and a millimeter long—sent fine wires to a titanium pedestal, about the diameter of a nickel, that protruded from Matt's head. When the skin healed, a fiber-optic cable would connect Matt to a computer that analyzed and stored his brain's electrical nerve signals.

If—and this was my clinical career's biggest *if*—the experiment succeeded, Matt would control a computer's cursor by only *thinking* about it, just as though he were moving a computer mouse with his paralyzed hand. That would remove a few links in the chains of his paralysis.

As a clinical investigator for the BrainGate study, I had discussed with Matt and his parents the many risks and potential benefits of the implant. Doctors, nurses, neuroscientists, and clinical engineers had planned every step and coordinated all the details, along with Cyberkinetics Neurotechnology Systems, Inc., the company formed by researchers at Brown University to develop the BrainGate. I remembered the research team's excitement as we worked on this revolutionary technology—but I also recalled moments of fear that our dreams for the BrainGate would end in disaster.

I told myself to calm down. Almost a decade of studies with monkeys had convinced the Food and Drug Administration (FDA) and three institutional review boards (IRBs) to approve a study of the first human BrainGate implant. Feeling less nervous, I reflected upon Matt's journey into disability and upon his yearning to overcome paralysis through the BrainGate, stem cell therapy, or any other technology.

A nurse dabbed Matt's face with a damp cloth. He was sweating in response to the stress of surgery.

"Please be careful near his nostrils," I said. "He's got MRSA; it's not that far from his nose to the incision."

The nurse looked taken aback, but it was clear she understood my fear that methicillin-resistant staphylococcus aureus—a variant of the benign bacterium we all harbor—might infect Matt's surgical wound. She moved the washcloth away and promised to pass on my concerns to the other nurses.

I left to find Matt's parents, Ellen and Pat Nagle, who were waiting downstairs. Standing near the elevator on the hospital's sixth floor, I stared out a large tinted window while familiar questions streamed into my brain with the sunlight. What did the BrainGate study mean for people with severe paralysis? Why had Matt subjected himself to this experiment, the first on a human being? Why had he, after having been paralyzed, undergone yet another metamorphosis—this one almost turning him into a cyborg? Why had his parents allowed him to do this? I also thought about my role in recruiting Matt for the study, which had many risks and only nebulous chances for success.

Led by John Donoghue, the chairman of Brown University's neurosciences department and a cofounder of Cyberkinetics, our research team had designed the study to answer the following questions:

1. Would the device be safe in human beings and not cause complications such as brain infections, bleeding, seizures, or any of the twenty-eight other adverse events listed on the informed consent form?

2. Would the implant record signals from single neurons, and do so reliably?

3. Could the BrainGate use these recordings to function as a neural prosthesis, like a bionic brain?

4. With the BrainGate, would a paralyzed patient be able to control a computer cursor through thoughts alone?

5. Finally, would the device allow advanced uses, such as controlling a television or opening e-mail? (The possibility of playing

video games had also excited Matt, but this wasn't mentioned
in the thick protocol submitted to the FDA.)

While training in rehabilitation medicine—also known as physical
medicine and rehabilitation or physiatry—I had developed an interest in
spinal cord rehabilitation. During and after my residency at Boston University, I treated many patients with spinal cord injuries due to car accidents, falls from buildings, dives into shallow pools, gunshots, ruptured
aortic aneurysms, bicycle accidents, and stabbings. Each of these patients
taught me about the complexities of this tragic condition that affects almost
every bodily system. They taught me about kidney failure, surgical bowel
complications, pneumonias that start out as colds, fevers from infections
or from staying out in the sun too long, festering pressure ulcers, urinary
infections that lead to septic shock and death, fractures in bones leached
out by paralysis, and depression—which led one of my patients to use his
car's hand controls to drive off the road and into nothingness.

What would Matt Nagle teach me?

Through the glass corridor, I saw a small maple tree in a cylindrical
white planter, its dark red leaves twirling in the summer breeze, and I
imagined its roots circling in search of new, nourishing soil that would
allow the imprisoned tree to grow taller and wider in the sun and the
wind. I stopped at the physicians' lounge to call my wife and give her a
quick update. Giselle, also a physician, had some misgivings about the
BrainGate project, but she supported my research activities. She was
relieved to hear about Matt's surgery and passed on the news to our children, Nita and Jacob, who had seen me on TV in connection with the
research and found the project fascinating.

At the café, two of my BrainGate colleagues were relaxing after the
surgery. I briefly sat down at the marble-top table and raised my coffee
cup in congratulations.

"Yes, the surgery was successful and the patient survived, but we'll
have to wait and see about the BrainGate," Burke Barrett said. As vice
president of clinical operations for Cyberkinetics, he was right to be
cautious.

"I feel good about the implant," Dr. Gerhard Friehs said. Still wearing
his green scrub suit, the lead neurosurgeon moved his head around to
relieve the tension in his neck.

"How soon can we test the system?" Burke asked, and sipped his coffee.

"Matt's young," Gerhard said, "so he should heal fast. But the drainage has to settle down. I'll let you know."

I left Burke and Gerhard to go outside and greet Ellen and Pat Nagle, who had been at the hospital since dawn.

"He's fine," I said. I pulled up an outdoor metal chair and joined them on the brick patio under a tall, shady beech. "Matt's a tough guy; his heart and lungs are in great shape."

"You know, he was one hell of an athlete," Pat said, closing his right hand into a fist and shaking it for emphasis.

"He sure put on a great performance today," I said.

"He'll never quit," Pat said. "Let me tell you about when he made MVP."

Matt's final high school game was against Walpole, which had a running back who went on to play college ball. The winning team would go to the Massachusetts Super Bowl, so Matt had to win. On the defensive line, he kept getting blocked, but then he got through just before the handoff to the star running back. Matt had him in his sights until his opponent did a little shake, and Matt got faked out. The star went right by him. Matt was knocked down while his opponent, eighty yards from the end zone, was on his way to a touchdown.

Matt jumped up and ran, pumping his arms and legs like pistons, breathing fast and shallow to get just enough oxygen to keep his muscles burning. He shocked the running back at the five-yard line and pulled him down. The crowd went wild, the Weymouth fans rising up in the stands to clap and yell and scream at the heart-racing defensive play. Even a few of the Walpole fans clapped surreptitiously. The other team rushed for the remaining yards, scored, and eventually won, but Matt's play was the game's highlight. A reporter's photograph showed Matt's back with a blurred "21" on his jersey, his left hand clenched in a fist, and his right leg charging forward as he reached for the running back.

For their performances, Matt and his buddy, Mike Romig, won the Weymouth Football MVP award and a thousand dollars to put toward college. But a game like that deserved a celebration, so they spent their money on kegs and food for a party with their many friends.

The Nagles were silent after Pat finished his story and came back to the present. Ellen stared down at her coffee and quietly said, "But it all ended three years ago, Independence Day. It all ended. . . ."

Until now, I had neither asked nor heard the details of what had happened to Matt on Wessagusset Beach in Weymouth. I brushed away a serrated leaf that had fluttered down onto the wrought-iron table.

"The thing that really gets me is that Mattie was hurt about three years ago, and I still can't talk about him without . . ."

Pat wiped away some tears and smiled ruefully. He no longer had the toughness of a cop built like a linebacker, the homicide sergeant who had cracked one of the most infamous murder cases in Cambridge.

"It's just horrible, a nightmare that doesn't end. I always thought I was stronger than that, you know?"

Ellen placed her hand on her husband's muscular forearm.

Under the dense green foliage, the Nagles described how Matt was severed from his life as a fun-loving star athlete, stripped of his strength, and humiliated by his basic bodily functions. Because of his condition, Matt cheered on researchers who were racing to treat severe disability with brain implants, electric limbs, or stem cells. These technologies gave rise to teams that were striving to treat human spinal cord injuries. At stake were millions of dollars, scientific careers, and academic prestige. Brain implants compensated for paralysis via an end run around the damaged site, the static line of scrimmage. Limbs could be implanted with electrodes that sent electric currents to stimulate muscle movement. Stem cells healed the body from within, through natural mechanisms of repair.

And, although completely paralyzed and no longer on the football field, Matt was also racing. He was in a race against time, a race to get a computerized brain implant, an electrode system, stem cells, or any other technology that could cure his spinal cord injury—before he died from its many complications. Based on his clinical status, Matt could expect to live to forty-five. With paralysis, sensory loss, abnormal movements, bowel and bladder incontinence, pain, and sexual dysfunction, Matt was one of the most disabled people I had ever met. Over the years, I talked on many occasions with Matt and his parents—as well as relatives, friends, doctors, caregivers, priests, teachers, coaches, Cyberkinetics researchers, electrode-implant manufacturers, stem cell scientists, and journalists—to understand and describe his story and his dream to regain movement, any movement, anywhere in his body, through any means.

The Hunting Knife at the Beach Party

Matt, July 3 to 15, 2001

On July 3, 2001, Matt went to the Fourth of July celebration at Wessa-gusset Beach, a popular inlet with a seawall to protect the adjoining road and houses. He wanted to watch the fireworks, have a few beers, and catch up with old friends. Before the roads to the beach closed, Matt and his friends gathered at a nearby house. This was an ideal way to avoid traffic and the cops, who prowled the beach because, every year, a fight or a drug deal or a car accident would occur. By evening, about three hundred people had danced onto the beach anyway, carrying their six-packs and boom boxes. It had rained, but the clouds were drifting away. The sand was still warm from the day's sunshine, and Matt enjoyed a Coors Light as the sunset bled through the clouds floating across the ocean's calm horizon.

He sat in Katie Perette's white Jeep Cherokee, the top off and the doors open to blast out techno-pop with a hypnotic dance beat. Many years earlier, they had been hockey teammates, and Katie still reminded Matt about his time in the penalty box. Before going after a player on the opposing team, Matt would say, "Katie, this one's for you," and whack her goalie pads with his hockey stick before speeding off on his skates. If they were behind, Matt would collect the team near the penalty box, hit the boards with his stick, and yell, "Let's go, let's go!"

Matt and Katie shared a beer; then he pulled her out of the Jeep for a dance. Blond and tanned, her long arms and midriff visible in a light summer shirt, her laughter harmonizing with the music, Katie looked really good without a padded hockey uniform. Why hadn't he seen this side of her until now? He had hung out and hooked up with so many girls, but he had always thought of Katie as just a friend—until now. He breathed a little faster, excited by the beer and the music, and said, "Hey, babe, let's go down to the beach for a walk."

"Get out of here, Matt. We're just buddies."

"Just you and me," Matt said, this time with an arm on her bare tanned shoulder. "Let's go down and watch the fireworks."

Katie knew Matt's penchant for intense but short relationships. She enjoyed his friendship and didn't want romantic complications, but she had second thoughts as she looked at his handsome face, framed with dark brown hair, and his body, athletic and graceful.

Before Katie had time to explain how fond she was of him, a fight broke out somewhere in the crowd and quickly turned into a brawl involving more than a hundred people. The Jeep's door was smashed in, a kid from Dorchester yelled for his boys, and someone shattered a beer bottle into a jagged weapon. A Weymouth kid ended up at the bottom of a pile and shouted, "Get Nagle!" because Matt always helped a buddy in a fight. Their friend Brian was bleeding from a stab wound underneath his eye, so Katie grabbed a towel to staunch the bleeding. She saw Matt dive into the pile of twisted bodies. He started pulling people off, just as he had done in his football days.

Then someone plunged an eight-inch hunting knife into the back of Matt's neck. The assailant ran away as Matt collapsed.

Katie saw Matt lying unconscious on top of a white picket fence, one leg extended and the other twisted backward. His hands twitched as if he were grasping for something. He was unresponsive to her cries. At first, she thought only a beer bottle had struck him, that it was only a gash. But when Brian lifted up Matt's neck, Katie saw that the white fence had a dark red stain and Matt's lips were a darker blue than his eyes. She screamed.

Two paramedics who happened to be at the party rushed over. One started CPR, rhythmically pressing on Matt's chest and breathing into his mouth.

"He's dead, no pulse—he's not even breathing!" he shouted.

The other paramedic ran to get emergency equipment and came back, gasping.

"OK, give him a strong breath, and I'll intubate."

The paramedic tilted Matt's head back, opened up his airway with a stainless-steel instrument, and slid in a plastic tube to hook him up to the ventilator that was on its way in the ambulance. The siren howled closer and closer.

"Stay with us," he kept saying. "Hang in there."

Katie sobbed as the paramedic blew air into Matt's lungs through the plastic tube.

Policemen on bikes arrived, and so did the ambulance, wailing through the crowd. The emergency technicians thought that Matt, his face blue from lack of oxygen, was dying. To protect his spinal cord, they put a hard collar on his neck before attaching his tracheal tube to oxygen.

As the ambulance left, Katie banged on the rear door, weeping. A helicopter flew Matt to Boston Medical Center. The cops cleared the beach.

||||||||||||||||||||||||||||||||||||||

Pat Nagle had gone to bed around nine that night, expecting to be paged. As the senior homicide detective for Cambridge, he knew that Fourth of July weekends were busy—shootings, stabbings, suicides—so he grabbed sleep whenever he could. One year a student had jumped off a roof at the Massachusetts Institute of Technology—then another suicide occurred near Harvard Square.

Over the years, Pat had developed a routine. He didn't keep a phone in the bedroom, to avoid answering it and then rolling over and going back to sleep. His pager was far away from the bed, on the dresser, to force him to get up, go downstairs, and call the station, wide awake.

Close to midnight, the phone rang and rang. He picked it up and heard the unfamiliar young voice of a rookie at the other end.

"Mr. Nagle."

His head still misted with sleep, Pat got a pen and pad to take notes about the victim, the hospital, and the suspects.

"Mr. Nagle," the voice said again, instead of, "Sergeant Nagle."

Pat was shaken out of his slumber. Oh my God, he thought. He looked up at the boys' bedroom doors.

"Is it Mike or Matt?"

"It's Matt. He's at Boston Medical Center." BMC was the primary trauma hospital in Boston.

"What happened?"

"He got hit with a bottle," the young cop said.

Pat's rapid breathing subsided. He thought, A bottle? How bad could that be?

"How's Matt?" Pat asked.

The kid kept his cool. "Mr. Nagle, you'd better go. Matt really needs you."

"I'll be right there."

Pat ran to get dressed. He hoped that Ellen would remain asleep, which she usually did when he went to work. He wanted to race over to the hospital and find out what had happened before telling her. But she sat up and asked him what was going on.

"Something happened to Matt." Pat choked up.

"Oh my God!" Ellen cried out.

They jumped into his cruiser and blue-lighted it to Boston Medical Center at eighty miles per hour.

||||||||||||||||||||||||||||||||||||

Ellen leaped out of the cruiser at the ER. Thinking that Pat was on duty, some of the staff greeted him casually.

"Hey, Sarge, how're you doing?" a security guard said.

Pat nodded and kept walking.

A nurse commented, "Busy night for you, huh?"

Usually he would joke around with the staff, but not tonight. No one was at the front desk, and Ellen screamed that she had to see Matt at once. Now the nurse realized that the weeping and nearly hysterical woman was with Pat.

Another nurse quietly said, "Your last name is Nagle, isn't it?" Then everyone knew he was there for his son.

The nurses gathered to put their arms around Pat and Ellen, then took them to a side room. Pat had watched this scene before—the tragedies of other parents. He knew Matt must be in bad shape. Ellen sank into a chair. Pat remained standing until the arrival of Dr. Erwin Hirsch, the head of the trauma unit.

"I'm so sorry. It's the worst injury I've ever seen." The tall gray-haired man paused and took a deep breath. "Your son was stabbed, and the spinal cord was severed. He's lucky to be alive. The knife had so much force that the point broke off in the neck. Just missed a major artery. We don't know if he's going to make it."

"Fully severed?" Pat asked. He remembered a spinal cord he'd seen in an autopsy. The bones of the spine protected the fragile bundle of nerves about the thickness of his thumb, but he knew a knife could find a small opening.

"Yes," the doctor said quietly. "I'm sorry."

Ellen's cries and sobs echoed through the small, dim room. "Can we see our son?"

Pat had seen such injuries. Recently, on his beat, a young man had been punished for a bad drug deal with a bullet in the neck, leaving him fully paralyzed. Pat knew that the millions of nerves that traveled to and from the brain, once damaged, could never be spliced back together, could never be repaired. In the course of his police work, he had heard of nerve reconnections in the arm or the leg, but never the spinal cord. Pat had consoled other families in similar tragedies, and he knew that this accident could be the end for Matt.

On the way to the ICU, Pat pulled the doctor aside, said that he was a homicide detective, and whispered, "I want to know what you think."

"It's not good," Dr. Hirsch said. "His chances aren't good."

The doctor did not tell them about the dismal history of these injuries. The surgeon who authored what is now known as the Edwin Smith Papyrus, an ancient Egyptian medical text, described the challenges of a spinal cord injury at the neck: paralysis of all four limbs, bowel and bladder incontinence, and sexual dysfunction. He applied meat and honey to the neck, but without optimism; in the classification scheme of Egyptian medicine, this was "an ailment not to be treated." Hippocrates began using traction for spinal patients, and his stretching apparatus was used as late as the nineteenth century. Prone and stabilized on a platform, the patient was pulled upward by the shoulders while the spinal deformity was pushed back into place. Traction can improve the alignment of the spinal column by helping the flow of nerve impulses, and it is still used on occasion.

In the second century CE, Galen's surgical experiments on animals proved that severing the spinal cord caused paralysis and sensory loss. A seventh-century Greek physician, Paulus of Aegina, developed the first effective surgical treatment: the spinal column was opened and explored to correct structural abnormalities. Today, this procedure is called a decompressive laminectomy. The Renaissance physician Ambroise Paré further refined this technique, but patients often died because of surgical complications. When faced with these risks, some physicians did not even attempt the surgery. After Lord Horatio Nelson was shot at the Battle of Trafalgar during the nineteenth-century Napoleonic Wars and lost "all power of motion and feeling," the ship's surgeon said, "My Lord, unhappily for our country, nothing can be done for you." Within a few hours, Nelson was dead.

With the introduction of ether and antiseptic techniques, surgery gave some hope for SCI. In the early twentieth century, Harvey Cushing developed criteria for surgical versus nonsurgical treatment. He believed that complete spinal injuries had a poor prognosis. A standard text of his time stated that the physician was doomed to "knowing that the patient was approaching an early death, to keep him alive for weeks and months on end, only to see him rapidly fade away, despite all skills and efforts."[1]

In the ICU, the Nagles were led past the nurses' station to room 10. Ellen closed her eyes briefly and said a silent prayer before entering the room with large glass windows. She had no idea what her son would look like, if Matt would even recognize them, or if he would survive the night.

Ellen didn't remember much of what the doctor had said, except the words "paralyzed" and "spinal cord."

Matt's eyes were closed, but he looked as handsome as ever. There was no blood, his hair was combed, and Ellen couldn't see the wound in the back of his neck. Dazed, she glanced around at the clean, bright, spacious room filled with medical equipment. A large, white, wheeled box with a computer screen had two plastic tubes leading into a smaller plastic tube that entered Matt's mouth and throat. He lay on a high-tech bed, hooked up to a monitor with green waves pulsing across the screen, indicating that his heart was strong as ever. Red and yellow numbers on the monitor tracked his vital signs.

Ellen went over to kiss Matt and tell him that they were there and that everything was going to be all right. But as she drew closer, Ellen knew it wasn't going to be all right, and she might lose her son. Matt's blue eyes, his best feature, were obscured by closed eyelids, and he didn't open them in response to Ellen's tearful endearments. Even if he had tried to talk, the plastic tube in his throat that connected him to the ventilator wouldn't have let him. A hard blue brace prevented movements of his neck, protecting his spinal cord. The red wound near his left ear was barely noticeable. An IV pump dripped clear saline from a large bag and a yellowish liquid, maybe an antibiotic, from a small bag with a white label. Ellen cried while Pat held her, stroking her blond hair as he murmured that Mattie would be OK. He was a fighter.

In search of hope, Ellen remembered Matt's ability to recover from injuries. As a one-year-old, he had explored, run around, and crawled on the stairs. When he fell and fractured his right leg, Matt had to wear a cast from his toes up to his thigh. But even with his splintered bones encased in plaster, Matt refused to slow down, dragging the cast behind him as he ran around or played basketball with his older brother, Mike, who went along with the fun. Three weeks after the fracture, the orthopedic doctor was surprised at the X-ray.

"Another week and we'll take off the cast." He tousled his young patient's hair. "But no more fractures, Matt."

Unsure of what the man in the white coat had said, Matt nodded and then showed how fast he could walk with the cast.

The doctor shook his head and smiled. "So that's how he was stimulating the bone growth!"

Ellen asked for a priest, but none was around so early in the morning. Sister Claire, who was covering for the chaplain, appeared out of nowhere, hugged each of them, and said, "God will take care of Matt."

But Pat was so angry that he responded, "Sister Claire, if God were a person, I'd punch him in the mouth."

"You know," she said calmly, "Matthew didn't die on the beach for a reason. God had other plans for him, and someday you'll know why."

Pat and Ellen kept a vigil as Matt struggled to survive, and they prayed that death would remain locked outside the intensive care unit. At sunrise, the ICU staff asked Pat to take Ellen home.

"You have to take care of yourselves so you can take care of Matt," a white-haired nurse said. "You're exhausted."

In his job, Pat had learned to appreciate insomnia, to keep going for the first crucial days of a murder case before it went cold. He wanted to work this case, to methodically analyze the crime scene and question the people who could tell him who had stabbed Matt and why. *Why?* It made no sense. Matt was an easygoing kid who liked hanging out and having a few beers with his friends. Pat Nagle's hazel eyes had not yet wept and wouldn't for a while. For now, the analytic part of his brain was in full control as he focused on what he would do once back in Weymouth. But he listened to the nurse for Ellen's sake, and he drove her home.

Now they had to tell Mike about his younger brother. Quiet and sensible, Mike hadn't even wanted to go to the beach the night before. He and his girlfriend had left early, unaware that the ambulance and police cruisers were there for Matt.

Mike was in the kitchen when the Nagles entered the house. As Ellen wept, Pat described Matt's condition.

Mike ran up to his brother's room, crying, "No, not Matt! Not Matt!"

He pounded repeatedly on the closet door until Pat came into the bedroom, grabbed Mike, and said, "Buddy, I can't hold you. You're six-four, and you can throw me all around this room, but I'm not going to let you go. I'm not."

Mike appeared to calm down, but he then wrenched himself free of his father's embrace and drove to their church. He ran up to the altar at Saint Francis Xavier, raised his fist, and screamed at God, "Why did you do this to Matt?"

Father Sean Connor, a young priest who had arrived at the parish only six weeks earlier, rushed out of his office.

"What's wrong, Mike?"

"Matt was stabbed," he said. "My brother might die." His voice shaking, Mike told the priest what his father had told him.

"He's still alive," Father Sean said as he gave Mike a hug. "And Boston Medical is the best trauma center around. Mike, there's still hope. Come—let's pray together for God's help."

The priest helped Mike kneel down and started saying, "Our Father in heaven. . . ." Father Sean explained the meaning of the last two lines of the Lord's Prayer in the context of this tragedy: "And lead us not into temptation, but deliver us from evil."

Fearing that Mike would try to find the attacker and also be harmed, Pat sped to the church in his cruiser, lights flashing. He found his older son kneeling in prayer with Father Sean and thanked the priest, who offered to help in any way possible.

Once Mike was safe at home, Ellen fell into bed. Pat tucked her in and went off to do his job: find the criminal. He took his gun but forgot his badge, a rare omission. In spite of Pat's grief, decades of experience kept him focused on the basic questions of detective work: Who had done this to his son? Why had the fight started? Had Matt started the fight, or had it been someone else?

Pat interviewed Matt's friends, starting with the kids who had been on the beach. He went from home to home, following every lead, tracking down every person who could help his search for the stabber. None of them knew how badly Mattie was hurt. They thought he had just been hit with a bottle.

And when Pat said, "It looks like Mattie might die," the typical response was one of shock. Pat told them what the doctors had explained to him: "If Matt lives, he's going to be paralyzed."

When Pat arrived at Tommy O'Neill's house, the state police were already there, following the same trail. They shook Pat's hand, tried to comfort him in the way of men who are used to injuries, mutilation, and death—until a friend or colleague is affected. Pat knew it was the worst thing in the world to investigate a personal case. But he was on his way to the projects on Lake Street and felt he was getting close.

As he headed toward the area, two detectives from his homicide unit called, asking to meet him at his house on Rustic Drive. Pat couldn't ignore this request; the boys had driven all the way from Cambridge to talk with him. At his home, Pat's fellow police officers promised him that everything possible was being done to find the stabber. Pat tried to explain what the doctors and nurses had said about spinal cord injuries. One of the men put his arm around Pat to guide him toward their unmarked cruiser, and they drove to Boston Medical Center. After escorting Pat to the seventh floor, they took off.

It dawned on Pat that he didn't have a car. He had been kidnapped and stranded—by his own men. A state trooper must have alerted the Cambridge police station about him. Once Pat got over his chagrin and even smiled at this trickery, he realized that this was best for everyone. He had to take care of Ellen, Mattie, and Mike.

Pat knew that his brother, Rick, a lieutenant detective with the Massachusetts State Police, was talking to the investigators but wasn't actively involved. Even though it was a long weekend, the detective bureau in Weymouth was working full-time. This was the law-enforcement brotherhood at its best, taking care of one of its own. Seeing the investigation in action was therapeutic for Pat.

America's Most Wanted host John Walsh, whom Pat had met at an international homicide conference, also offered to help. Pat was well-known in police circles for solving the murder of Jeffrey Curley, a ten-year-old boy abducted by pedophiles, and the case had been similar to that of Walsh's own son. Walsh was ready to do a show to generate public attention and help catch the stabber. But he was soon identified: a twenty-year-old carpenter who had a criminal record of assault. He turned himself in at a police station in New Hampshire and was sent back to Weymouth, where he was arrested and placed in jail.

⁓

In the ICU, Matt struggled to wake up. The lights were dim. Something kept flashing on a screen. Had the sun already set? Where were the stars? What was that machine near his bed? What was that beeping? An alarm clock? Was it morning already? Why wasn't he on the beach? An insistent, high-pitched sound. A woman in a white uniform. Where was Katie? What was that other woman doing here?

The woman adjusted a machine attached to a metal pole, which had a plastic bag dangling from a hook. The beeping stopped. She left. What a strange bed! He was resting on nothing, suspended in midair. This place smelled nothing like the beach, no salt air. Bleach. Who would bleach the air?

Matt slowly turned his head to the other side. A bright room where people dressed in white sat at a large desk. Some wrote, others talked on the phone, and one stared at a row of computer screens. Greeting cards were tacked on to a bulletin board in front of him. Flower vases and ornamental potted plants were on a small table. There had to be an explanation.

Matt yelled out—but he heard nothing. His throat was sore, not like he had a cold but worse, as though someone had stuck a finger down his mouth. He raised his right hand to wave at the people in white. Nothing happened. Then Matt tried his left arm. Nothing. He lifted up his head. Were his arms strapped down? No. A white blanket covered his body. He kicked up his right leg to get someone's attention. Nothing. His left leg. Still nothing. Was he chained down? Was he in a prison? Had someone paralyzed him with drugs? Was it the clear liquid dripping into his right arm from the plastic bag? His arms and legs lay helpless before his eyes. He tried to get up, but only his head moved to the side. He tried until his neck started to ache. He felt nothing below his neck. Nothing.

Matt looked around his cage, which was full of machines. The largest one was a box on wheels, with knobs and dials and a white plastic tube snaking toward him, into his throat. Why was it puffing away, regular as a person breathing? Just above his head, another machine was connected to his chest with wires; it blinked out red numbers. What was it trying to tell him? Had he turned into a machine, some kind of cyborg?

Then, gradually, Matt remembered the fight, the pile of bodies, his friend at the bottom, the sharp neck pain. On the warm sand of the beach, he had felt waves of moonlight crashing over his inert body, drowning him. He remembered that he had struggled awake and coughed the darkness out of his lungs. Then he saw a line of people going up a flight of ivory stairs, moving slowly in a procession through a kind of hazy barrier toward a brilliant, nurturing light. He also tried to climb the stairs, but no matter how hard he tore at the barrier, he couldn't get through. Above the ocean, white seagulls circled around the light, and he heard human and bird voices speaking the name of God. But he still couldn't get through and gave up as he stared longingly at the light and envied the people on the great staircase. And the waves of moonlight kept crashing over him, but he swam through them and reached the shore—then everything went black.

That was all he remembered. He still didn't know what was going on. He thrashed his head from side to side, his eyes darting wildly from machine to machine as they stared back at him.

The woman in white came back and saw him looking around. She dashed out to the waiting room and returned with his mother. Ellen told him that he had been stabbed in the neck and was completely paralyzed. He was at Boston Medical Center. It was the fifteenth of July.

Matt was silenced by the ventilator that breathed for his paralyzed muscles, but he frowned, his eyes widening, as if to say, *That's impossible!*

His mother wept and showed him the *Patriot Ledger* article with the headline PARALYZED WEYMOUTH FOOTBALL STAR FIGHTS ON. Matt read the article that his mother held up for him. When he came to the point where someone was arrested for stabbing him, he understood. Paralyzed. Not able to move. Ever again.

His inaudible scream flew out of the ICU room, silent waves rippling through the entire medical center and beyond, over Boston Harbor and out into the ocean.

Every nerve that traveled from Matt's brain cortex to his spinal cord had been severed, disconnected from his once-powerful muscles. Normally, upper motor neurons travel down in the spinal cord to the neck and back; there they connect with and control lower motor neurons that move the arms and the legs. Each level of the spinal cord also receives sensory information from the peripheral nerves. Messages of pain or touch ascend to the brain in the sensory-nerve bundles of the spinal cord. Almost every single neuron that sent sensory information to Matt's brain had been cut. He had sensation only at and above his shoulders.

The major nerve pathways to and from the brain are located in the white matter of the spinal cord. Myelin sheaths, like white insulation on electrical wires, result in fast nerve impulses: 120 meters per second, or 270 miles per hour. One can get a sense of these speeds by driving on a highway at the speed limit—then imagining moving five times faster. A cross section of the unmyelinated gray matter is shaped like a butterfly, and these nerve impulses move slowly (see figure 2.1). The motor neurons in Matt's brain were still intact, but the messages to his spinal cord went from flashing speeds to zero when they hit the wall of the knifepoint. If Matt had been stabbed just two inches below his neck, he would have had normal function of his powerful arms.

Each square millimeter of brain cortex contains about one hundred thousand neurons, or a total of around fifteen billion neurons. To visualize the cortex, one might look at the outer three-millimeter rind of a ripened cantaloupe. But Matt's brain could neither send motor signals to his spinal cord nor receive sensory stimuli from his body.

Ellen knew that Matt wasn't used to losing. He always assumed that he could fix whatever was wrong, push himself to overcome barriers, or race past an opponent. As he glared at the spinal cord injury from across the scrimmage line, he wouldn't believe that it was unbeatable.

Ellen and Pat felt the full impact of Matt's injury on their own lives. Almost finished with her master's degree at Lesley University, Ellen wanted to become a reading specialist. While teaching kindergarten, she had discovered the challenges some children face and wanted to help them and their teachers with her specialized training. Now her education was on hold so she could help her own disabled son.

Late every night, after Ellen had spent a draining day at Boston Medical Center, Pat watched over Matt. After Ellen took some medicine for sleep, Pat rested with her for a while, then quietly got up, took his gun, and went to the hospital. At the ICU, he watched Matt breathe with his machine until dawn. When Pat got home and crawled into bed, Ellen was still in her drugged sleep, unaware that he had been away.

Crowds of well-wishers came to the Nagle house on Rustic Drive, blocking traffic for the neighbors, who never complained. People whom Ellen and Pat hadn't seen for years brought casseroles, money, and sympathy. The Nagles were grateful for all their support but had to tell the story again and again. Each time, Ellen would cry. And then they'd tell it again.

Some of the visitors reminisced about Matt's athletic ability, which they cited as a reason to be optimistic about his recovery. Coach Paul Naper recalled nine-year-old Matt's determination to excel at hockey. He was always the first one on the ice, his skates fresh from their weekly sharpening, an oversize shirt covering his hockey pads and his slim but muscular body (see figure 2.2). Young Matt glided onto the ice, pushing off his right leg, then his left, his speed picking up as he headed straight for a crash into the boards. The coach almost covered his eyes and prepared his first-aid bag—until Matt scraped the ice and stopped, a few inches from injuring himself. Then the coach realized Matt was practicing the art of legally checking an opponent, pushing him into the boards to knock him down or blast the wind out of him long enough to get an advantage and perhaps score a goal. Matt's next drill was to weave from side to side in a zigzag pattern, cutting through imaginary members of the opposing team and guiding the puck with his stick, which was almost as tall as he was, until he could pass it to an imaginary teammate. When Matt slowed down, his rapid breaths condensed in the cold air and wrapped around his face.

Coach Naper's team had a successful and busy season—sometimes they played four games per weekend. Matt's skill, enthusiasm, and team spirit made him the unofficial captain. On New Year's Day 1988, his team reached the championship game of the Santa Claus Tournament in Hingham and skated away to a 4–1 lead; but Dorchester tied the game, sending it into overtime. As play progressed, each team had to reduce its players on the ice until there was only one skater and the goalie. Overtime play was harder and more intense than the regulation game, as there were no teammates for support.

When Coach Naper looked over at Matt, he nodded, and the coach smiled. At the next line change, with only one skater allowed per team, the coach put Matt back on the ice. Matt had already made a name for himself in the town of Weymouth, and a chant rose from the crowd: "Matt Nagle! Matt Nagle!" When the puck dropped, Matt leaned into his opponent, twisted the puck away with his hockey stick, raced past him, swung once to fake a slap shot, swung again after the goalie lunged mistakenly, and sent the puck flying into the top open corner of the goal. The play took only ten seconds—the Dorchester goaltender didn't have a chance. Matt had scored for the second time in the game, but this goal won the tournament. The place went wild, and Coach Naper thought that the crowd's roar would blow the roof right off the rink.

Pat took a leave of absence from supervising the homicide unit in the Cambridge Police Department. He was used to functioning on a few hours of sleep and catnaps, but he couldn't keep up his nightly vigils with Matt in addition to his work demands.

Pat and Ellen watched their son fight one complication after another. Being on a ventilator in the ICU increased the risk of infections, and Matt struggled with a pneumonia that attacked three lobes of his lungs. Then he fought off a urinary infection. Both conditions caused high fevers, and, once, his temperature rose to 105 degrees.

Ten days after the stabbing, two of Matt's friends, Meaghan Murphy and Brian Hemenway (nicknamed Hemma), were allowed to visit. Meaghan felt scared and nervous on her way up Interstate 93 to Boston Medical Center. How could this have happened to one of their best friends?

Meaghan's favorite memory of Matt was his invitation to play with the Rottweiler puppies at the Nagle house. He knew that she loved dogs, so he brought her into the Nagles' garage, which doubled as a kennel for the five pups and their mother. Matt played with them regularly so they would develop friendly dispositions; Meaghan remembered how they crawled all over him, climbing up to lick his face and neck.

Now, at the modern, multistory Boston Medical Center, near the older brick medical school, a tall cherry tree was almost bare of its pink blossoms, and the nearby bushes were clotted with violet hydrangeas. Meaghan and Hemma stopped at the security desk inside the lobby, which was brightened by large skylights and bamboo palms reaching for the glass ceiling.

Inside the ICU, a dimly lit cave with beeping machines, Pat thanked them for coming and gave a status report: Matt was in tough shape and was still disoriented. Their presence was important, but they also had to be strong for him. Tears streamed down Meaghan's face as she listened to Pat. Then he looked directly at her.

"If you don't think you can handle it, you shouldn't go in there."

She nodded and wiped away the tears. But when she went into Matt's room, she was taken aback at seeing Matt helpless, wires and tubes connecting him to machines.

Matt looked up at Hemma for a moment, then at Meaghan. He tried to speak, but the tube in his neck tore up the sounds, and they couldn't understand him. Even if he had been able to speak, they might not have understood him; he seemed confused, his eyes randomly scanning the small, crowded room. At other times, though, he seemed lucid, as if he were attempting to convey something very important.

They struggled to understand, but nothing made sense. Meaghan and Hemma repeated, "What, Mattie? What's the matter?"

Matt stared up at the ceiling, his eyes wide open, and started crying. Despair struck him, and there was nothing he could do to defend himself.

After a few more unintelligible murmurs and choked cries, Matt started looking back and forth between Meaghan and Hemma. They finally figured out that he wanted Hemma on the left side of the metal bed and Meaghan on the right—to push him out of the ICU.

Meaghan and Hemma looked at each other. Meaghan gave a barely visible, sad smile.

"No, we can't take you out of here, bud," Hemma said.

Matt got upset and tried talking through his tube again, but soon he gave up. Meaghan kissed him on the forehead; then she and Hemma left, both crying as they walked out of the ICU.

⁓

Two weeks after the stabbing, Matt whispered hoarsely through the tube that someone was trying to get in through the seventh-floor window to hurt him. At first Pat thought this was a conspiracy. Maybe a friend of the stabber was trying to kill Matt to keep him from talking about the crime. Then Pat learned that the pain meds and tranquilizers were causing hallucinations; perhaps Matt was also having nightmares.

Some days, Pat would wake up and ask himself if Matt's spinal cord injury was just a nightmare. Then, one morning, when Pat was in bed, hoping to salvage a little sleep from a restless night, the front door opened, and he heard Matt say, "Dad, Dad!"

He jumped out of bed, ran downstairs, and yelled, "Matt!" But there was no sign of his son.

iiiiiiiiiiiiiiiiiiiiiiiiiiiiiiiiiiiiiiii

It would be a miracle if Matt survived. A few of the ICU staff suggested that it would be understandable if Pat and Ellen chose to pull the plug, to give up on Matt.

"You might want to consider a different approach," a gray-haired nurse said to Ellen with a kind smile. She must have seen terrible injuries in the trauma ICU, some leading to severe disabilities that ruined families emotionally, physically, and financially.

The Nagles did think about it when Matt was intubated and sedated. They wondered how long Matt had been without oxygen, blue, almost dead after the stabbing—and whether his brain was damaged. Later, when pneumonia nearly killed Matt, one of the ICU doctors asked if they wanted to disconnect the ventilator.

How could they even think about these questions? How could they end this precious life they had brought into the world? Ellen talked to an ICU doctor who said that Matt was essentially a "head on a bed." How cruel that sounded! But Matt would always be connected to a breathing machine, and what kind of quality of life was that? They heard about the inevitable hardships and the heartaches of caring for Matt, the insurmountable obstacles if they chose to keep him alive.

Ellen asked herself if she would allow him to suffer. She didn't even know if Matt was in pain. What did they know for certain? And what did Matt want? She had heard that some people with complete paralysis didn't want to go on, but there were also people like Christopher Reeve who never gave in to their paralysis.

One of the most helpful people at Boston Medical was a young doctor who specialized in rehabilitation medicine. She gave encouragement and practical information, such as how to communicate with Matt in spite of the trach tube in his throat. This doctor had seen people like Matt survive and go on to lead productive and even happy lives. Ellen would always be grateful to her.

Sister Claire also became very important to Pat and Ellen—their guardian angel. Every time they went to see Matt, she was there with a hug or a kind word. Now Pat understood her comment about the reason Matt had not died on the beach. If Matt had died, Pat would have lost both Ellen and Mike. Neither of them would have been able to cope.

Pat lived in a fog of grief, able to see only a short distance ahead, uncertain if he would trip and fall. There were times when he felt neither happy nor sad, when he felt nothing except the mechanical drive to keep moving through the next step, and the next step, and the next. For days at a time, Pat slept very little, as if he didn't need to rest, but his face appeared drawn with grief.

Sister Claire must have talked to a doctor, who pulled him aside and said, "You have to sleep. It'll regenerate your body. And you have to be in shape to take care of Matt."

"I can't, I just can't," Pat replied. Although he tried again and again to force his brain to descend into deep sleep, he couldn't. Then, another ICU family, whose son had a brain injury from a car accident, suggested that he try Excedrin PM. Pat avoided medications; but he had to try something, and it worked.

While Ellen was at Matt's bedside, Pat stepped aside to ask Dr. Hirsch about his son's prognosis. He didn't think Ellen was ready to hear the cold truth. They stared at each other outside the ICU.

"I want to know," Pat said. "Just tell me what you think."

The trauma surgeon took him to the small side room, where they had first met.

"This is probably as good as it gets. From his neck down, he's lost everything. He's dependent on the vent." He paused, then added, "He'll never walk again."

Pat stared out the window at the Boston skyline. "OK, Doc."

There was nothing to discuss. But even with that prognosis, it didn't make any difference. They still had Matt. He was still part of their lives. His brain seemed to be fine, and he was still Matt. Pat didn't know how they would have reacted if he were brain-dead, but luckily they didn't have to even think about that.

Then Ellen came in and said, "I want to know what's going on with my son."

Pat realized he had to be honest with his wife, and he told her what he had just heard. Ellen sat down and started sobbing.

The door to the side room burst open, and an elderly woman came in. She was weeping. The small area became filled with loud sniffling noises as she dabbed at her eyes with a tissue. Then she slumped into a chair and stared at a pastel painting of lilies. Pat and Ellen went over to comfort the grieving mother. She had been wrestling with whether to stop life support and donate the organs of her comatose son, whose small car had been hit by a pickup truck.

The doctor offered to talk to them later, then quietly shut the door.

More than two weeks after being stabbed, Matt burst through the offensive line of complications. His fever passed, and he was weaned off the stronger pain medications and tranquilizers. Soon it became clear that Matt would not be able to breathe on his own, because his chest muscles and diaphragm were also paralyzed. Near the end of July, Matt had surgery for a tracheostomy, a hole in his throat for a plastic tube that would be connected to a respirator—probably for the rest of his life. A one-way valve would allow air to pass inside but not go out through the plastic tube, instead sending the air through the vocal cords to allow Matt to produce speech. Ellen and Pat knew this procedure was necessary, but they were also excited because, at last, Matt could talk to them.

The Nagles could only think of their son's determination and his spirit. Both felt that Matt still had too much to give and too much to live for. They chose to keep Matt alive, even if on a respirator.

But Matt also had to make that decision. Ellen couldn't bring herself to talk to him about it, but Pat thought he could balance his own despair with optimism for his son, tempering his own grief with the joy of still having Matt alive.

Ellen and Pat also heard about the promise of stem cells for repairing Matt's spinal cord. A doctor mentioned a study about researchers who transplanted stem cells from bone marrow into the damaged spinal cords of rats. Tissue analysis showed that the cells changed into specialized neural cells that wrapped a reparative sheath around the injured neurons. Maybe there was hope.

A few days after the trach tube was surgically placed in Matt's throat, Pat told his son about the severity of his injuries. First, he gently patted

Matt's cheek, for he knew he would feel his touch there. He told Matt about everything: the paralysis, the breathing on a ventilator, the loss of almost all feeling, the loss of his bowel and bladder control, the loss of his sexual function . . . the loss. Matt closed his eyes and cried silently.

"Buddy, if you want to fight it, we'll fight it together—as a family."

Pat squeezed Matt's shoulder, as he had sometimes done when their hockey team was losing.

"Mattie, the doctors said that if you choose to die, they'll disconnect you from the respirator. They'd do that for you. The bottom line is we'd let them—for you."

Pat swallowed and turned away so that Matt wouldn't see the tears squeezing out of his eyes. He wanted Matt to make his own decision, but he couldn't help adding, "We believe in God and hope and stem cells. And we believe in you."

The silence of ICU room 10 was broken only by the regular hissing and puffing of the ventilator, the white noise of death. Pat Nagle waited for his son's response. He would sit there as long as Matt wanted to think about his condition and his future—until the next day or the next week or the next month. And he would honor his son's decision.

Still crying and with beads of sweat on his forehead, Matt opened his eyes. Through pain, through the white plastic tube that stifled his voice, he gasped out, word by word, "Dad . . . I . . . want . . . to . . . fight . . . this."

3

Stairway to Recovery

Anna Iacono

Anna Iacono rested under a white blanket in the hospital bed, the right side of her face sagging, her right arm and leg completely paralyzed.

"Good morning," I said to my new seventy-five-year-old patient.

On February 19, 2011, a blood clot in Anna's left middle cerebral artery shut off the blood supply to almost half her brain, leaving her nerves starved for oxygen and glucose. Like circuit breakers, her neurons stopped their flow of electricity while her brain used its remaining nourishment to preserve as many cells as possible. But as parts of her left brain lost almost all blood supply, her neurons began to die, and she had a stroke.

An MRI scan showed damage in the frontal, parietal, and temporal lobes as well as structures below the cortex, deeper in the brain. Each of these areas serves specific functions, depending on whether the right or left brain is damaged. Having read her MRI report, I feared for her sensory function, swallowing, memory, organization, and planning. Anna's right hemiplegia, or complete paralysis on that side, was sadly obvious. Arm function is controlled by nerves on the brain's surface, and leg movement by nerves deep within the brain, so I knew that Anna had a large stroke. (Figure 3.1 in the illustration insert shows the brain's structure in relation to motor function as a homunculus, which comes from Latin and means "little human.") But she still had her left side and would be much less disabled than people with paralysis of all limbs, like Matt.

Before patients are admitted to my center, I review their records to see if they are appropriate for inpatient rehabilitation, where every day they receive three hours of physical, occupational, and speech therapy. So I already knew that Anna needed a lot of help with basic activities, such as moving around in bed, sitting up, going from her bed to a chair, bathing, dressing, using the toilet, and grooming. Her awareness of her right side was limited, so I stood on her left for our initial interaction. (During her stay, the rehabilitation team approached her from the right, to improve her awareness of the affected side.)

Her speech was a jumble of words, but her brown eyes spoke clearly to me, expressing uncertainty. After explaining my role, I asked a few simple questions to determine the extent and nature of her aphasia, which is a term for speech impairment.

"Is your name Anna?"

"No," Anna said, shaking her head, and now I noticed that her short blond hair was remarkably well coiffed for someone in the hospital. Then she corrected herself. "Yes."

I thought, Good, she can correct herself after producing the wrong answer, but is it reliable? To check her accuracy, I said, "Is your name Fred?"

"No, no, no," Anna replied, frowning at me as if I were demented.

Well, that's a good sign, I thought. And I liked that emphatic response from a strong-minded woman.

"Do you know what happened to you?" I asked.

"I had a—" She pointed at her head with her left hand.

"You had a . . . ," I said, looking quizzically at Anna, trying to get her to finish the sentence.

"I had a so, a so," she said.

"Yes, it was a very bad stroke, but you'll get better. Your therapists and nurses and I will help you recover."

Anna gave a small smile, as if now daring to hope, but the right side of her mouth didn't rise up like the left.

As for her speech, the syntax was preserved, but the vocabulary was disrupted. I pointed to my tie and asked what it was called, a standard question that yields a variety of answers from patients with strokes and aphasia. Some give approximations like "tay," while others offer related phrases like "in a dress." The most affected patients say unrelated words or phrases, such as "I water."

Anna stared at me, uncomprehending. When I again pointed at my tie and asked if it was a book, she responded with "tux." Now I felt optimistic about the prognosis for recovering her speech. At least "tux" was accurate in its first consonant and its relation to clothing.

As I examined her, Anna understood my directions only half the time due to her aphasia, so I demonstrated the neurologic tests. I squeezed my eyes shut, gestured for her to do the same, and said, "Now you try that." This approach helped. Then I looked around in all directions and stuck out my tongue at her. (I once had to explain the reason for this test of the twelfth cranial nerve to a family member who walked in just as I was showing an aphasic patient how to protrude his tongue.)

On her own, Anna could not move her right shoulder, elbow, wrist, or hand. Perhaps she would do better with my help, so I tried to activate her muscles by lifting up her arm, forearm, hand, and fingers. There was no movement. Then I asked her to move without the pull of gravity, by supporting her arm and forearm as she tried to bend the elbow. Finally, for each joint, I lifted her upper limb and dropped it into my other hand, hoping for some resistance to the force of gravity. There was no muscle activity.

When my examination of her arm was finished, I forced a smile, held her small, thin hand, and said, "It's still early, so there's hope." Anna nodded.

I supported her right lower limb with both my hands and found only trace movement at the hip. Not even a muscle twitch was present at her knee or ankle. Her left side, however, was quite strong for an elderly woman. She would be able to do a lot with half her body. I noticed that she was slim yet muscular, another good sign, as she would have to rely on half her strength to propel her entire body.

Because of her inability to communicate, I could not reliably assess Anna's sensation to light touch. She didn't feel the pressure of my fingernail on her wrist or ankle; I was discouraged by this finding but said nothing to Anna. She probably had a damaged thalamus, which is the sensory relay station of the brain. I tapped some muscle tendons for reflexes. Finally, I checked her muscle tone by moving her right arm and leg back and forth; her limbs were flaccid, another bad sign for recovery.

One of the first questions that people ask me after a stroke is, "Will I walk again?" Had Anna been able to utter that question, I would have replied that I was hopeful. I am always optimistic, for often that is all we have in rehabilitation medicine. But, to be entirely honest, I would have said that walking was possible but unlikely. Even for skillful and experienced therapists, retraining Anna to walk would be a challenge. In addition to her flaccid right side, she had to contend with reduced sensory awareness; it is difficult to control a limb if a patient is not aware of its position in space. And her limited communication skills could be a barrier to working with the therapists.

On finishing my examination, I placed Anna's right hand on her abdomen, so that she would be able to see it. Anna and I looked into each other's eyes.

"You've had a very bad stroke," I said, "but we'll help you get better." I smiled and touched her right shoulder, but she did not seem to notice. When I offered my left hand, she reached out with her own left hand. Anna tried to smile, but it was difficult to force up the right side of her mouth. Instead, she nodded her head a few times. I nodded back. In this crucial first interaction, I felt that we believed in each other—she in my role as her rehabilitation doctor, and I in her will to recover as much as her damaged brain would allow. Now, as I looked into her eyes, I felt that her expression had changed from uncertainty to trust.

Later, I spoke to Anna's older sister, Vinnie (short for Vincenza), to find out what had happened. Also a blond, Vinnie was dressed in an argyle sweater and black slacks. Her eyes were puffy and reddened. Vinnie told me that they had enjoyed a typical Sunday on the day of the stroke, with a trip to the hairdresser and then shopping at Macy's.

"Anna likes to dress to the nines—high-heel shoes, pressed pants, Ralph Lauren sweaters, diamond earrings, gold bracelets." Vinnie glanced at her sister, now wearing a blue hospital gown that covered her like a sheet, and sighed. "We went to Walmart for a while and then came home. Anna took a nap in front of the TV."

Vinnie paused, as if something had just occurred to her.

"You know," she continued, "Ginger never left her side that night. Usually our cat sits with me. Anyway, Anna stayed up late, so I asked if everything was OK. She didn't say anything, just looked at me. I wanted to call 911, but Anna tried to knock the phone out of my hand." Vinnie shook her head and said that her sister was always a strong-minded person.

Vinnie told me that Anna had been widowed for thirteen years and that they had lived for the past seven years in a condominium in northern Rhode Island, with a view of the Stillwater Reservoir. For the past forty years, Anna had worked as a buyer for a drugstore and liquor store, where she also made gift baskets. In fact, she was still working full-time.

"Anna does all the cooking, dishes like lasagna or spaghetti and meatballs—Italian food. She really likes making Easter pies for the family." Vinnie turned away and wiped her eyes with her sleeve. "I guess she won't be doing that this year."

"Well, let's hope," I said, although I was afraid that Vinnie was right. Later, Anna's daughter, Deborah, told me that her lemon meringue pies and cream puffs were heavenly, and I said that I felt the same way about my late mother's pecan pies. I did not mention that my mother was born about the same time as Anna. There is always the risk of premature death after a stroke.

Then we talked about whether Anna would ever walk again.

"She'll definitely need a brace," I said. "I don't know if it'll be a short plastic one for her ankle or if her knee joint will also need support. It's too early to say."

A short brace is ideal because it forces the muscles of the knee joint to work, which leads to better recovery. But if Anna's knee muscles did not regain any function, she might need a brace that encased her from the foot to the thigh, immobilizing her ankle and knee as if they were in a cast. Anna's leg paralysis also made me worry about blood clots, which typically form in the leg and travel up the thigh, sometimes breaking off and going to the lungs. In the second to fourth weeks after a stroke, the most common cause of death is a pulmonary embolus, a blood clot in the lungs. Because of some bleeding in her brain and the recent use of an intravenous blood thinner, I could not use another anticoagulant; that could worsen the bleeding and possibly cause her death. So I prescribed sleeves that were intermittently inflated by a pump, which squeezed her legs and helped the flow of blood. I also resolved to carefully watch her for any signs of clots.

During the course of inpatient rehabilitation, I learn a great deal about my patients. Commenting on the photographs and cards in a patient's room is one of the best ways to start a conversation. Vinnie told me that Anna had six grandchildren and had recently welcomed her first great-grandchild, Penelope, whose photo was central in a gallery on her bulletin board.

"Very cute," I said, pointing at the photo. "I didn't know that was your great-granddaughter."

Anna's mouth widened in a smile, and her right side was almost symmetric with her left.

Later that day, I learned from Anna's physical therapist, Colleen Fitzsimmons, that Anna had no movement on her right except for some trunk muscles, which help with balance. Anna could sit at the edge of her hospital bed with support from her left arm for only fifteen seconds before toppling over to her right. Colleen had to do most of the work for Anna when she sat up and moved from the bed to the chair. In the therapy gym, Anna could not stand up in the parallel bars.

Colleen's therapeutic plan was to determine Anna's abilities, her potential to recover, and the ways in which physical therapy would improve her condition. During the first week of Anna's admission, I observed her with Colleen, a tall, athletic therapist with advanced training in neurodevelopmental therapy (NDT). NDT is based on the clinical practice and research of Berta Bobath, who developed this treatment approach with her neurologist husband, Karel, in the 1940s. While providing physical therapy to children with cerebral palsy, she realized that their patterns of movement were abnormal. Bobath carefully observed these awkward movements; then she designed treatment protocols for specific problems such as weakness or hyperactive tone in certain muscle groups. Her husband helped her with a theoretical framework based on the normal development of children. For instance, infants start off with simple movements—they turn their heads, then their trunks. At first, they have so-called primitive reflexes, such as the startle reflex, in which a loud noise makes a child fling out the arms and bend the legs. Another example is the asymmetric neck reflex; when the head is turned to one side, the child extends the arm and leg on the same side, like a fencer.

In her physical therapy, Bobath tried to reduce muscle tone and the primitive reflexes that impair smooth and functional movements. Through tactile stimuli and positioning, she elicited the mature postural responses that normally replace the primitive ones of infancy. To counteract the stiff and jerky movements caused by cerebral palsy, she developed patterns of movement that were fluid and graceful.

Later on, Bobath realized that adults with strokes had dysfunctional movement patterns similar to those of her young patients. Both conditions involve damage to the upper motor neuron that originates in the highest level of brain function, the cerebral cortex. Starting in the 1960s, the Bobaths applied NDT to rehabilitation for adults with strokes who were hemiplegic.[1] NDT is consistent with the concept of neural plasticity, the foundation of rehabilitation therapies: with natural and repetitive movements of the patient's body, the brain's neurons will change and adapt to neurologic conditions such as strokes. Research has shown the brain's remarkable ability to change the organization and function of nerves; through rehabilitation therapies, existing nerve pathways are healed, and new connections are formed. For example, Anna was encouraged to repetitively advance her weak leg in a fluid manner and was discouraged from spasmodic movements, although both were effective in walking.

With Colleen's help, Anna would have to go through the process of human development again. While Anna lay on a blue foam mat on a therapy platform, Colleen placed her in different positions on both her left and right sides. Berta Bobath had described three stages—flaccid, spastic, and relative recovery—based on the patient's muscle tone. Initially, Colleen knew that she had to help Anna bear weight on her weakened right side, which would encourage the return of muscle movement in normal patterns. To retrain Anna to walk, Colleen would support the trunk on one side for better posture while advancing her patient's leg with her own. Once muscle tone returned (in the spastic stage), Colleen would prepare Anna for the final phase of recovery; this stage involved exercises for muscle groups like the ones that extend the knee (quadriceps) while walking. By providing both sensory and motor input, the therapist helps the patient regain normal movement.

Anna's therapy had to start at a basic level—on the therapy platform in the gym. Colleen assessed her patient's abilities with movements that most of us take for granted but that are laborious after a stroke. Anna's balance was tested while sitting upright, crouching on all fours, kneeling, and standing. Colleen also analyzed her patient's abnormal movement patterns and weakness in specific muscles. This evaluation established a baseline for monitoring her progress and planning her therapeutic program.

For instance, Colleen noticed that Anna tended to lean toward her unaffected side, a common posture after a stroke. This occurred because of sensory deficits: she was not fully aware of her right side. As a result, her balance was off. Anna's communication was also limited, so the challenge was even greater.

Five days after Anna's admission, I was surprised to learn that she was able to do most of the work of moving from her bed to the wheelchair. She could stand while holding on to the wooden rail that lined the corridor, but she had to sit down in her wheelchair after a few seconds. How had she made such progress in only a few days?

When I examined Anna, she had only slight movement of her thigh while seated. Lying on her side with my hands holding her leg up in the air, Anna could barely move her thigh, and there was only a muscle flicker at her knee. When I gently dropped her leg on the bed, it was limp and had no muscle tone, a bad sign for recovery. So how was Anna activating her paralyzed leg muscles in order to stand? To answer this question, I decided to observe her during physical therapy that afternoon.

In the gym, I waved at Anna, who waved back with her left hand. She was wearing a burnt-orange sweatshirt over a white shirt and black pants with a matching orange stripe. Earlier, I had seen her struggle with a brush to adjust her short blond hair. Her light pink lipstick was perfectly applied except for a missing area at her right lower lip. I knelt down, supported her leg, and tried to move it back and forth to help with her movements.

"Try and kick it up."

She strained to kick out her right leg, but I felt only a twitch of her quadriceps.

"It making meld," Anna said. For an hour each day, she worked with her speech therapist, Miranda Scott, who was pleased with her progress, even though it seemed slow to Anna.

I interpreted her comment to mean that she was improving and said, "Good, keep going; your knee's getting stronger."

"Then it's biddle," she said.

"Your strength is coming back." Certain that "biddle" meant "better," I patted her right shoulder.

"Yes." Anna grasped my hand with her left hand.

Colleen helped Anna lie down on the therapy platform, with her right leg hanging down.

"Pull, pull, pull," Colleen said as Anna tried to activate her hip and knee muscles. Once Anna was lying on her back, Colleen asked her to push up her torso to form a bridge. At first, Anna had not been able to even lift her right pelvis.

"Ready? Push up—tight, tight," Colleen said, encouraging Anna to activate her trunk muscles. A strong core is essential to movement of the limbs. Once Anna succeeded, Colleen said, "And down—slowly, slowly. Good." Exercising a muscle by tightening it is important; letting it relax in a slow, controlled manner also helps build up strength.

Moving farther away from the trunk, Colleen worked on Anna's hip rotators as she rolled her leg in and out repetitively. Then it was time to exercise Anna's right knee. Colleen supported the limb, ready to assist whatever movement her patient could muster.

"OK, kick up, kick up." Colleen also tapped Anna's quadriceps muscle in order to facilitate its movement.

Anna's son, Bill, also watched as Colleen tapped the front of Anna's thigh and said, "Kick, kick, kick. Hold it, hold it." She turned to Bill and said, "I'm beating up your mother."

We shared a laugh, and then Colleen asked Anna to slowly let her leg fall down as her quadriceps relaxed.

While considering Anna's impressive progress, I decided to learn more about her, to understand the source of her mental strength. Vinnie told me that they had season tickets to the Providence Performing Arts Center, where Anna had enjoyed *The Phantom of the Opera* and *The Lion King*. They also liked the movies, and I wasn't surprised to learn that Anna especially liked comedies.

"She has a positive outlook on life," I said.

"Yes, she enjoys life," Vinnie said. Occasionally they took a bus to New York to shop and see a play. "We'd planned a trip in June, but I don't know if that'll happen."

A week after her admission, I saw Anna in the gym with Colleen and her occupational therapist, Liz Lima. To work on her balance, Anna kneeled on a blue foam mat on the floor while Liz held up colored plastic cones and said, "OK, nice and slow, come forward." She was trying to get Anna to reach out of her base of support, to activate her trunk muscles.

"Slow, slow," Liz said, as Anna leaned forward and almost fell into Colleen's arms.

Then Colleen started an exercise in which Anna knelt in front of her on the mat and they moved sideways. Colleen occasionally helped her slide over and offered encouragement. "Pick it up," she said. "Yeah."

If Anna became tired, Colleen helped by picking up her right lower extremity. Her posture also drooped, and Liz asked her to "stay tall." When Anna breathed rapidly and slowed down her movements, Colleen said, "OK, take a rest. That was a lot of work."

Later that day, Anna sat at the edge of the platform, and Colleen asked her to repeatedly lift her right leg off the ground and let it down. On occasion she said, "Pull, pull, pull." Eventually Anna was able to do this by herself, but she was clearly fatigued.

"Nice job," Colleen said.

Even at this early stage, I felt more optimistic about Anna's recovery. She was determined to get better, even if the therapy was exhausting.

Ten days after Anna's admission, I saw my patient and her physical therapist in the hallway. Colleen asked her to move up in the wheelchair and bend forward to prepare to stand up. She placed Anna's left arm on the wheelchair so she could use all available muscle force to stand up, as she had to rely on only half her muscle strength to move her entire body.

"One, two, three, up," Colleen said as she lifted Anna up while holding on to her trunk. As Anna rose up, her right leg rotated out and a plastic brace supported her foot. Although neurodevelopmental therapy avoids the use of such braces, they are often necessary to help the patient regain mobility. Without a brace, Anna would drag her foot, increasing her risk of falls. On Anna's left hand, I saw the veins popping out; they were visible through her fine, fair skin as she held on to the handle of the hemi-walker. This assistive device has four legs that are angled away from the patient, who places the device at the side instead of in front. A hemi-walker increases the patient's base of support and improves stability. Its name derives from the fact that it is ideal for a person with hemiplegia.

I observed another therapy session two weeks after Anna's admission. In addition to the plastic brace, Colleen applied an air splint around

Anna's knee and inflated the clear plastic sleeve to prevent Anna's knee from buckling. Now Anna was ready to walk a good distance.

"How does it feel?" I asked.

"Swell, that goes into back," Anna said.

I shook my head to let her know I did not understand. She nodded, as if realizing what had happened. She sat on the therapy mat, with its height adjusted to three feet, and leaned forward, in a hurry to stand up. Her eagerness was partly due to impulsiveness, which can occur after a stroke, but mostly because of enthusiasm. I have rarely seen such determination. In view of her severe paralysis, Anna's attitude was even more impressive. But her weakened condition made her vulnerable to falling.

"Slow down," Colleen laughed. "You give me agita."

She bent her patient's knees to position them for standing up and then placed Anna's hand on the hemi-walker. With Colleen's hands guiding Anna, the therapist and patient moved effortlessly, an unlikely melding of the tall, young, dark-haired therapist with her shorter, elderly, blond patient. I smiled at the contrasting pair and stayed to watch the rest of the therapy session.

Colleen held on to the left side of Anna's waist as she bent forward at her patient's right side. Both the air splint and the leg brace stabilized her lower leg, which would have collapsed otherwise.

Colleen said, "Walker," and Anna advanced the device with her left hand. "Right leg," said Colleen, and Anna moved her thigh forward. She and Colleen moved as one unit to advance the right leg. Then Colleen said, "Left," and Anna advanced her left leg.

The sequence went on: Anna first advanced the walker on the left leg, then the right, then the left, again and again, slowly but surely, with Colleen's assistance to help her straighten up and avoid falling. Anna walked a distance of twenty feet down to the end of the hallway, where I checked her leg, which was clearly more fatigued but still had some strength at the knee.

Her partial muscle recovery led me to suggest a trial of electrical stimulation, which I thought might reeducate her muscles and help her regain movement. When Colleen tried this therapy, however, Anna experienced hypersensitivity due to her stroke and could not tolerate the stimulation, which her damaged brain perceived as painful. This abnormal sensation was more evidence of a compromised thalamus, the brain's sensory center.

Two days later, I asked Anna, "How's therapy?"

"It's finally better, not as it was nine work."

"How is it better?" I said, unsure about her second phrase.

"Got very bull and claw." Anna vigorously nodded to emphasize her comment, as if she knew that her words were still garbled.

"Do you have any pain in the leg?" I asked, trying to make sure there weren't any problems. At times I just did not understand my aphasic patient, but it was important to continue the dialogue. That would help with recovery of her speech.

"I got bell sundry," Anna said.

"How's the walking?" A great deal of interpretation is necessary when talking to a person with aphasia. Perhaps "bell" meant "better," and there was no leg pain.

"They did wally with over there." She pointed to the gym. Then she pointed at her right leg and said, "It goes down the ways."

"It'll get better," I said.

"OK." Anna gave a thumbs-up to show that things were improving.

||||||||||||||||||||||||||||||||||||||

Anna's remarkable progress kept me wondering why she was so successful with her rehabilitation. Part of the reason was that she was slim and had to move around less body weight—with only half her strength. Her cognitive function seemed relatively well preserved, considering the size of her stroke, so she was able to understand the therapists and follow their lead. Then I learned from Vinnie that, three times each week, Anna would get up at 4:30 AM and go to a nearby gym called Elite Fitness. She also stayed active by gardening in the summer, when she planted a circular plot near their condominium with tulips, lilies, and daffodils.

Almost three weeks after Anna's arrival, I saw her walking with Colleen, who was occasionally applying pressure at her flank for better posture and more effective movements of the right leg. Anna no longer required the air cast at the knee, but she still needed the plastic leg brace and the hemi-walker. Colleen told me that Anna had been glad to work at her therapy whenever possible. Once I was surprised to learn that

Anna had done six hours of physical, occupational, and speech therapy in one day. Patients are expected to do three hours each day, an exhausting schedule after a stroke, and I often find them resting in the early evening.

"Was it hard or easy?" I asked at the end of her walk.

"Actually, it was a billing room for me," Anna replied. Her damaged brain was trying to produce words that were similar to what she wanted to say. Anna must have hoped that her team of clinicians would be able to interpret her speech, for she kept trying to communicate.

"Anna's exercise is giving me big muscles," Colleen said.

Anna smiled at her physical therapist's comment, then advanced her hemi-walker, occasionally dragging her right toe on the ground. Some therapists might have given more assistance, but Colleen didn't. Anna strained to swing her right leg forward. This act made Anna shift her weight onto her left side and more easily advance her right leg. Colleen held on to the waist of Anna's sweatpants and applied pressure at her right lateral chest to facilitate her trunk flexor muscles while offering encouragement.

"Kick it. Thatta girl," she said. As they neared the wheelchair, Colleen said, "OK, let's finish strong."

Anna marched forward, advancing her hemi-walker on the left, then her right leg, then her left leg.

"How did it feel?" I asked her.

"I feel like, like a . . . feether."

"You didn't feel light as a feather to me!" Colleen laughed and pretended to wipe her fatigued brow.

I imagined that Anna now felt relatively light as she walked. At first, she had almost no movement and probably felt like a mass of flesh that had to be lifted from her wheelchair by the rehabilitation staff. And as Anna recovered from aphasia, her words sounded more and more like what was intended.

"It looked like hard work," I said.

"Nope, go there keep going," Anna said.

"Yes, keep going," Colleen said, and the two walked on down the hallway.

While chatting with Deborah, I learned that, three years earlier, Anna had gone to Italy with her granddaughter, Victoria. Having recently returned from a trip to Italy, I thought she would enjoy reminiscing. I described entering the Florence Cathedral via a small, dark foyer, which led into an enormous interior with Gothic arches, an inlaid-marble floor rippling out in concentric patterns, and a floor plan in the shape of a cross. When I looked up, I saw that the dome was decorated with stunning frescoes of *The Last Judgment*, choirs of angels, and Christ and Mary with the saints. At the bottom of the cupola was a fresco of the deadly sins and hell.

A smile flickered on Anna's face as she heard my descriptions, which could hardly do justice to the magnificent cathedral. Occasionally she nodded, as if remembering details from her trip to the Florentine duomo with her granddaughter. I recalled admiring the stained-glass windows. Those in the aisles were of saints from both Testaments, and the circular windows in the dome depicted Christ and Mary. The windows had been created by some of the greatest Florentine artists, such as Donatello. I told Anna about the narrow stone staircase that spiraled up to the balcony near the base of the dome, where we had a panoramic view of the city.

"The guide said we had climbed 463 steps, and it sure felt like it," I said. "But it was a small price to pay for the view of Florence at sunset. The best part was sharing it with my son, Jacob."

Anna smiled, patted my hand, and nodded as if she knew exactly what I meant.

A few days later, I saw Anna again walking with Colleen, but now she was using a four-pronged quad cane instead of the hemi-walker, because her balance had improved. She could walk up to fifty feet at a time.

After she landed in a wheelchair at the end of her walk, I asked how it felt.

"It feels like moss squiggle," Anna said.

Her right leg appeared shaky, and I assumed she was referring to her instability. I said, "The squiggles will get better."

Anna smiled at me and nodded.

The following day, when a nurse's aide saw Anna walking in the hallway, she started doing the electric slide and said, "You'll be dancing soon, too."

Anna smiled, nodded her head, and said, "Yes, I am."

Her daughter told me that Anna and her husband had loved to go dancing at a dance hall, called the Farm, in southern Rhode Island. The couple's favorite dance had been the jitterbug, and they had especially liked the song "Mack the Knife."

"You went a long distance," I said. "How did it feel?"

"It felt like a nettle person," Anna said.

"Better?"

Anna nodded yes.

"How's the brace working?"

"I got half this?" Anna said.

"Yes, you need it—for now."

"I hope." Anna tapped her knee and adjusted the brace.

"Keep it up." I interpreted her aphasic speech and said, "I also hope you won't need it someday."

"I'm sure to best," Anna said.

<hr />

On March 22, a month after Anna arrived for rehabilitation, her balance and strength had improved to the point that Colleen felt she was ready for stairs. She was walking with the plastic brace, which supported her foot and was not uncomfortable, so the device was ideal for her. (In contrast to Anna, patients discussed in later chapters on stroke rehabilitation required high-tech solutions for their lower limb weakness.)

The center's gym has a wooden staircase for physical therapy, with standard dimensions for each step as well as two rails (see figure 3.2). This challenge was far easier than the stone steps of the Florentine cathedral. Anna held on to a rail with her left hand—her entire right upper extremity was still paralyzed—while Colleen stood on the right, helping her patient's balance as she first lifted up her left leg, then her right. Colleen reminded her of the mnemonic for climbing stairs: "Up with the good." Her foot still dragged, and she needed the plastic brace, but Anna cleared the steps most of the time with minimal help from Colleen.

As she lifted herself up the steps, I thought of how much energy Anna had to expend compared to a person without paralysis. I thought of how much mental strength was required for her to climb a few steps, a routine task for most people. I thought of Anna's remarkable recovery from almost complete paralysis, and I imagined her once again in Florence, walking along cobblestoned streets, ascending to the top of the marble-floored cathedral.

Alive in the Electric Chair

Matt, July to September 2001

A month after his injury, Matt received an envelope from the Weymouth Police Department. Pat wondered what was coming back to haunt Matt. Too many parking tickets? Prepared for another bad surprise, Pat opened the letter and was immediately taken aback. The letter informed Matt that his score on the entrance examination for new cadets was 97.

So the rascal had listened to him after all but hadn't said a word—typical Matt. Pat reread the letter, and his chest filled with pride. But the feeling vanished like air from a punctured balloon as Pat realized that his son would never follow the beat of his father and grandfather. Pat left the letter on the kitchen table for Ellen and went upstairs to their bedroom to lie down and stare at the white ceiling.

Many friends and family members visited Matt and supported the Nagles with phone calls, food, cards, flowers, letters, visits, and prayers. On a website for Matt, people left messages of encouragement and hope. With the help of Ellen's sister Lee Ann, the Nagles created a foundation to serve people with severe disabilities.

Among the visitors was a coach who reminisced about Matt's early football career. Concerned about potential injuries, Pat had discouraged football, but Matt had insisted. His speed, agility, and competitive spirit convinced the coach that he was ready to play, but there was one obstacle: the minimum weight requirement of fifty-five pounds. Matt weighed just over fifty, but with rocks in his pockets and the toe of the coach tipping the scale, he qualified. Then he proved that he could play with the bigger boys, running, tackling, and throwing the ball as well as anyone (see figure 4.1).

When the coach finished, he told other football stories. Matt and his neighbor, Jared Gray, often played touch football on Rustic Drive. When their older brothers heard the yells and commentary on every brilliant play, they joined in. Matt and Joey Gray teamed up against Jared and Mike Nagle. The street was the field; the curbs were out of bounds. The boys ran patterns for each other, pretending to be NFL players who raced through imaginary defenders. One of their heroes was New England Patriots quarterback Steve Grogan, who had passed for almost twenty-five thousand yards. The boys took turns pretending to reach that mark, with whoops and yells and imaginary media fanfare. For hours after school, they stayed out until nightfall.

When summer was over, Matt started playing football in a Pop Warner league, which prepared kids for playing in middle school. He got to know Mike Romig, who would be one of his best competitors and also one of his best friends. Both were gifted athletes with strong egos; both wanted to play quarterback. Street football had trained Matt to fight for the role of first-string quarterback. At the beginning of the season, Matt was tougher than usual as he proved himself in the new league, but he later regretted his initial hostility. Matt could tell that Mike was frustrated, and after the competition for quarterback had been settled, he went up to him and said, "That was a good fight."

In the football league, Matt played quarterback and also lined up beside Mike Romig on the defensive line. Matt had a reputation for hitting hard, despite his being one of the smaller players on the field. He helped his teammates regardless of which position he played, and after a pileup, Matt would get right in and rip his opponents off the stack until all his orange-shirted teammates were standing upright.

Some of the nurses at Boston Medical Center became very fond of Matt, who charmed them with his endearing smile and wink. One of the senior nurses, who had treated many patients with tetraplegia, told the Nagles, "You've got to get him to a good rehab facility."

Matt and his parents started preparing for a long war against the spinal cord injury. Ellen and Pat researched catastrophic-care hospitals. There were some fine rehabilitation units within local hospitals but no specialized center in Boston for people with complete paralysis who were also ventilator dependent. From talking to nurses and doctors and searching the Internet, they gathered information on rehabilitation hospitals in Denver, Atlanta, Miami, and New York.

Until World War II, rehabilitation for people with spinal cord injuries was essentially nonexistent. To care for soldiers with paralysis, England created specialized centers. In the postwar years, Ludwig Guttmann, a German-Jewish physician who escaped the Holocaust, pioneered a synthesis of medical and surgical treatment combined with physical and psychosocial rehabilitation. Later knighted for his research and his devotion to these patients, Guttmann was aware of the "thousands of years old prejudice towards spinal cord sufferers," but he was determined to "rescue these men and women from the human scrap heap and to . . . give them a purpose in life."[1] Guttmann developed essential strategies to prevent skin breakdown and bladder infections—which were often fatal complications—such as frequent pressure relief and intermittent bladder catheterization. His hospital offered vocational rehabilitation, special adaptive equipment, and even wheelchair sports. Guttmann's work remains a model for centers around the world, including the one where Matt received his rehabilitation.

"What's the *best* place for spinal cord rehab?" the Nagles asked all the clinicians.

The answer was always Shepherd Center in Atlanta, but Pat learned that the facility couldn't take Matt for three months. Other places had shorter waits, but Pat wanted the finest care for his son, so he started calling people. He got the names of priests in Atlanta from the Archdiocese of Boston, the names of police officers from his colleagues, the names of politicians from Massachusetts legislators—anyone who could help Matt. Pat called in old and almost forgotten favors, and eventually he expedited Matt's admission to Shepherd Center.

Matt left Boston in a Learjet, his aerial ICU. The Nagles followed on a plane provided by Angel Flight, a nonprofit organization of pilots and other volunteers dedicated to helping patients and families. At the entrance to Shepherd Center, the Nagles were greeted by a large bronze statue of a wheelchair athlete, his left arm outstretched for balance, his torso twisted for maximum force, and his right arm hurling a javelin. Now Ellen and Pat were certain that this was the right place for Matt. A large stone slab rising into the air described the center as A CATASTROPHIC CARE HOSPITAL. Also inscribed on the slab was the facility's logo, a flower blossoming within two cupped hands. Shepherd Center stretched out horizontally, a structure of alternating layers of brown stone and glass that had a spacious lawn with pathways for people in wheelchairs. Now the Nagles hoped that their son would be cured.

To assist out-of-town families, Shepherd Center provided small apartments to them at no cost and within walking distance of the hospital. Everything was taken care of so that the Nagles could focus on helping Matt recover. They saw it as a place of hope, especially after some medical professionals in Boston had raised doubts in their minds. There was no question of euthanasia at this world-class facility. Pat and Ellen were also glad that Emory University Hospital was nearby, in case Matt had an emergency.

As soon as he arrived via ambulance from the Atlanta airport, Matt began the arduous process of rehabilitation. Four nurses carefully turned Matt to his side and placed a long, white plastic board under him. After counting to three—and while making sure not to pull on his ventilator tube, IVs, and bladder catheter—they slid him over from the black stretcher. Matt's bed, designed for people at high risk of pressure ulcers, had balloon sections with alternating levels of air pressure.

Once he was comfortable, the ventilator was working, and his vital signs were stable, Matt received one visitor after another. His rehabilitation doctor asked a series of questions to review all his bodily systems.

"Do you have any movement or sensation below the shoulders?"

"No," Matt gasped out through the tube.

"Do you know if you have to urinate?"

"No."

"How about bowel movements?"

"No."

"Are you able to take a breath if the vent is disconnected?"

"No."

"Not even one breath?"

Matt shook his head.

"Are you able to swallow normally and eat and drink without choking?"

"No."

"Do you have spasms in your arms and legs?"

Matt looked down at his legs, which twitched helplessly at times. "Mostly the legs," he gasped through the ventilator. "They shoot out; I can't stop them."

"Ever have pounding headaches, sweating, and high blood pressure?"

"Yes, I've had dysreflexia."

"We'll watch out for that," the doctor said. "Any tightness when the therapists move your legs?"

"No."

"Let me know if that happens. Abnormal bone formation is a big problem for some patients with tetraplegia. I hope you don't get it."

"Me too."

"Any pain?"

"Terrible shoulder pain, burning and stabbing, like a hot knife."

"Sorry to hear that. We'll do our best to get that under control. Sometimes we have to use a combination of meds."

"Anything that helps. Some nights I can't sleep because of pain."

"Where is it the worst?"

"Where I lose all the feeling—like my body's trying to feel something again, but it's only pain."

"I think we'll be able to help you." The doctor reached down to gently pat his shoulder. "Any depression or anxiety?"

"Both. Bad." Matt shut his eyes, tightly, as if to choke off any tears.

"I'm sorry." The doctor adjusted his glasses. "I have to ask if you feel suicidal."

"Sometimes."

"Do you have a plan for . . . for ending yourself?"

"No."

"Good. Do you promise that you'll talk to one of us if the depression gets worse?"

Matt stared at the ventilator. "Yes."

"And is it OK if I get a psychiatrist to see you?"

"I'll talk to a shrink."

When the doctor finished his examination, he confirmed that Matt was completely paralyzed, had no sensation below his shoulders, and had abnormally brisk reflexes.

As the doctor was turning to go, Matt said, "Doc, will I ever walk again?"

"I'm afraid you have a complete injury. That knife slashed every single nerve in your spinal cord. There's no connection between the brain and the spinal cord. The knifepoint is still embedded in your spinal cord. That's not good for recovery either." The doctor looked down at the floor and then at the ventilator before looking Matt in the eye. "I'm terribly sorry. I don't think you'll be able to walk again."

Matt's physical therapist, Diane Johnston, came in after the doctor left and the nurse emptied his urine bag. Diane gave Matt a quiet smile and carefully checked for any movement.

"Try to bend your arm as hard as you can, Matt."

"I'm trying. Nothing happens," Matt said.

"OK, let's try the legs." The therapist found not even a flicker of a movement.

Matt saw her fill out a chart of his muscle testing. She wrote a zero at the top of each column—one for his right side and one for his left—and she drew a vertical line down each to show that the rest of the muscles were also fully paralyzed.

Cynthia Skinner, Matt's occupational therapist, focused on his arms, and she confirmed the absence of even a muscle twitch, which might have been used for adaptive technology.

So far, Matt had received little hope. But his speech and swallow therapist was confident he would eat real food again.

"We'll check your swallow mechanism with food and liquids of different textures. It tastes horrible, but the barium lets us videotape the food and drink as they go down. We'll make sure it doesn't go into your lungs."

"Hey, anything for a cheeseburger."

Matt started to salivate at the prospect of real food, which he hadn't tasted for a month. Due to fears that his damaged swallow mechanism would lead to pneumonia, he had received only tube feedings.

Even when patients have severely limited strength and movement, function is possible. People with spinal cord injuries at the upper neck (fifth cervical, or C-5 level) can raise their arms with the deltoid muscles and bend their elbows with the biceps. They can feed and groom themselves independently with adaptive equipment, but they depend on others for dressing, bathing, and toileting. They can drive manual wheelchairs with special rims and even specially adapted vans. Injuries at a lower level of the neck allow wrist extension and grasping; these patients have greater independence. Paraplegia occurs with injuries to the back, and, though these people are in wheelchairs, the full use of their arms allows complete independence.

But Matt had no movement of the arms. Nothing.

When Matt first tried to sit up—after a month of bed rest—his room grew dim, and he sank below the light as his blood pressure dropped. He awoke to the sight of his doctor and a small crowd of nurses and therapists. His bed was tilted, with his head down to get the blood back to his brain, and an IV flowed into his vein.

"Don't worry," his doctor said. "Everything's under control. We'll use elastic stockings and an abdominal binder to maintain your pressure. And the therapists will sit you up gradually on a tilt table."

Matt's long confinement to bed had robbed his muscles and blood vessels of their tone. And the knife had cut off his brain's control of his sympathetic nervous system, which helps regulate blood pressure. Left to its own devices, his sympathetic nerves could be overactive, causing sweating and headaches due to spikes in blood pressure—a condition called autonomic dysreflexia. If his sympathetic nerves were underactive, his pressure would drop and the blood would rush away from his brain, making him dizzy. It was tough for the former football star—who once did a hundred sit-ups at a time—just to sit up.

Once Matt could sit upright in a wheelchair, his speech and swallow therapist fed him oatmeal and liquids laced with barium. Matt had to prove that he could eat regular food without choking, coughing, or aspiration (which occurs when food goes down into one's trachea instead of the esophagus). Matt joked that his only aspiration was to eat real food. His therapist used a towel as a bib. In spite of her careful feeding, there was some spillage, and she wiped the white material off his mouth and chin.

When the video fluoroscopy was finished, Matt could tell by the smile on his therapist's face that he had passed. After the radiologist's blessing, Matt enjoyed his first real food in a month. Now, at least he had regained the basic human function of eating. Matt soon advanced from a soft diet to regular food, and one day, when the swallow therapist gave the green light, Pat rushed out for a cheeseburger. It was the best meal Matt had ever had.

A week later, Matt and his parents had a special session with his rehabilitation team for the family-teaching program.

"Family teaching?" Pat asked. "Aren't we going to try some kind of special rehab to get him back on his feet?"

"He's a complete injury," a nurse said, gently placing her hand on his back. "We're going to teach you how to take care of your son."

That was when the meaning of "catastrophic hospital" hit Pat. He felt as though he had gone up for a rebound, lost the ball, and landed so badly that he broke both ankles.

First, the nurses and therapists taught Ellen and Pat how to care for Matt's skin, to prevent ulcers that could get infected and possibly kill him if the bacteria spread throughout his body. Matt had to be turned side to back to side at least every two hours for pressure relief. He also needed a special cushion for his wheelchair, and he had to be lifted up periodically to prevent excessive pressure on his skin.

Ellen was also shocked at the implications of the family-teaching program. From the beginning, when Pat started calling everyone who could help Matt get into Shepherd, they had believed that Matt would somehow walk again. And they were committed to bringing him back home.

The Nagles learned about Matt's bowel and bladder program from Heidi, a cheerful young nurse from the Boston area who always had a smile for her fellow Red Sox fan. Matt had lost control of his bowels, just like an infant. But his situation was complicated because pain medications and other drugs affected his GI tract. As a result, Matt needed a stool softener and a medication that pulled water into his intestines to give his stool the proper consistency. Having lost the normal movement of his bowels, he required a stimulating medication. For severe constipation, Matt's options were a rectal suppository, an enema, or even disimpaction (the removal of stool by a nurse). If matters worsened, Matt might have to drink a bottle of magnesium citrate, which would have a laxative effect. For a bowel obstruction, he might need surgery. Ideally, Matt would have his bowel program every other day; a caregiver would stimulate his rectum to start a reflexive bowel movement. These scheduled bowel movements would eliminate the humiliation of incontinence.

As Ellen heard all this advice and watched Heidi demonstrate on Matt the proper setup and techniques, she flashed back twenty years to when her boy weighed 180 pounds less and she had placed him on the changing table, with diaper wipes to clean him and ointment to prevent diaper rash. Ellen smiled at the memory, then felt sad. But she remained calm, jotting down some notes while Heidi spoke. As for Matt, he said little during these sessions.

Next, the Nagles learned how to manage his bladder. Matt's brain no longer sent signals to his bladder from the urination center in the frontal lobe. Normally, when the bladder is full, it contracts, and the sphincter around its outlet relaxes—at a convenient and socially appropriate time. But Matt's bladder contracted while the outlet was still tightly closed, which sent urine up to his kidneys and could have led to renal failure. A bladder catheter is only a temporizing solution, for it increases the risk of urinary tract infections, widespread infection, and death. The best approach is to antiseptically empty the bladder with a rubber catheter every four to six hours. This should be done religiously to avoid infections—a common problem after spinal cord injuries—and kidney failure.

Matt had always been such an active baby, Ellen remembered, always trying to crawl around the house. Now he was completely immobile, unable to move in bed, let alone get out of it—a prisoner behind steel rails. Diane, his physical therapist, showed Pat and Ellen how to use the lift to get him from one place to another. Once in the electric wheelchair, Matt could at least go outside his room. His parents even learned the basics of wheelchair maintenance.

Matt's occupational therapist, who was in charge of rehabilitation for his basic daily activities, such as bathing, dressing, and toileting, also educated his parents. Cynthia emphasized the importance of getting Matt fully dressed in the morning so he would feel like a person, not a disability.

Skin care, bowel and bladder function, mobility, feeding, grooming, bathing, dressing, toileting, autonomic dysreflexia, spasticity, pain management, depression—the Nagles received a crash course in each of these areas. Soon Ellen's notebooks were filled with practical information about the management of spinal cord injuries, and their days were filled with classes about how to best care for Matt. Most important, the Nagles learned about the psychosocial damage of a spinal cord injury—the discouragement, the depression, the social isolation, the despair, the risk of suicide. Exhausted at night, they looked forward to their sleep, though it only lasted until six the next morning. When they felt ready, Ellen and Pat provided all of his care: feeding, grooming, bathing, dressing, and toileting.

In August 2001, Matt and his parents attended a conference with his clinical team to discuss his status, rehabilitation plan, and prognosis.

"I'm not sure how the knife blade missed the vertebral artery," the doctor said, pointing at the red tube going through a model spinal column. "Your injury's complete. Barring a miracle, you're not likely to recover the ability to breathe, move your arms, or move your legs. You'll be dependent for your daily activities and your bowel and bladder program. On the positive side, you could learn a lot and direct your care. And you can work on your communication. When you go home, you're going to need a lot of help."

The doctor's pager went off; he glanced down and ignored it.

Ellen stared at the tiled floor, stunned. Her athletic, handsome son was condemned to a ventilator, unable to breathe, unable to move, and unable to control his bowels and bladder. *Unable.* She brought herself back to the meeting and wiped away Matt's tears.

"Did you want to ask the doctor about your trach?" Ellen said.

Matt forced out a question in high-pitched whispers, but the doctor did not understand him, so Matt spoke one word: "scarring."

In the background, Pat blew his nose and said nothing.

"We'll get an ENT to check that," the doctor said, referring to an ear, nose, and throat specialist.

"Matt's having a real tough time with anxiety," Pat said. "If a door slams or there's a loud sound, his stomach gets jittery."

"Our psychiatrist will reassess your meds. And I'll talk to you later on."

Matt tried speaking through his trach. Ellen deciphered his whispers.

"Matt's got a painful cavity in a molar," she said.

Matt could be sent to the dentist's office with his ventilator. For now, the doctor offered more pain medications.

"Is there any way, down the line, he could get off the ventilator?" Pat asked.

"Unlikely," the doctor replied.

"Unlikely," Pat echoed softly.

A phone in the room rang repeatedly until someone lifted the receiver and quickly placed it back in its cradle to avoid interrupting the discussion. But there was little left to say, so the Nagles thanked Matt's rehabilitation team and wheeled him back to his room.

Matt was much stronger than Ellen and Pat, and he cried only for a little while. Then he asked for two of his favorite movies from Blockbuster. *Things to Do in Denver When You're Dead* portrays a hit man with a contract out on him. With the tables turned, he knows he will die over the weekend but continues living as he always did, even starting a romantic relationship. Another movie they watched was *Pay It Forward*, in which a twelve-year-old boy realizes that a good deed can never really be paid back. His solution to this dilemma is to be kind to someone else—in essence, paying the kindness forward—so that eventually the number of good deeds in the world will grow exponentially. Pat also rented a movie starring some attractive girls, which Matt had enjoyed in the past, but he didn't want to watch it.

"It's just too depressing," Ellen said when they were back in their little apartment.

"He was a ladies' man," Pat said. "A buddy told me Matt could have dated any girl in high school. I said we were so proud, but he didn't get the sarcasm."

"It's too bad he didn't want to make any commitments." Ellen knew the odds of a grandchild from Matt were slim because his injury could also affect his fertility.

"Why *would* he, with all these girls after him?"

They both smiled at the memory of his shenanigans.

<hr />

When Matt's medical status had stabilized and his blood pressure no longer betrayed him, his physical therapist, nurse, and respiratory therapist tried weaning him off the ventilator. Diane made sure his lungs were clear and asked Matt if he felt ready.

"Sure," Matt gasped out between the vent's forced breaths.

An oxygen tank was connected to a rubber bag, ready to manually force air into Matt's lungs in case of problems. The respiratory therapist disconnected the vent and placed a plastic cap on the white plastic tube protruding from Matt's throat. Matt had taken the first step in the ritual of ventilator weaning: corking the trach.

"Now you're on your own, buddy," the respiratory therapist said.

"Good luck, Matt," Heidi said.

Diane started a stopwatch, and the small group gathered around Matt for this experiment in his survival. While they stared at the oxygen level, at the color of his fingernail beds, and at his expression, Matt tried to take a breath on his own. He felt nothing below his shoulders, so he had no idea what his chest muscles were doing. Soon, he felt that something was wrong, as though he were running as fast as he could but not moving, as if on a treadmill.

At one minute and thirty seconds, the respiratory therapist said, "Sorry, Matt. Got to hook you back up. Your oxygen's down to 90 percent, and I can't take any chances."

Matt closed his eyes and said nothing.

The next day, Matt again tried to escape the ventilator, and this time he lasted one minute and forty seconds. A few days later, he reached a ceiling of two minutes.

"I can do longer, I can do longer," Matt said to his rehabilitation team and his parents.

"We really believed you'd come off the vent from sheer willpower," Ellen said.

"Buddy, you tried, and tried, and tried," Pat said. "But you just couldn't breathe."

Matt's doctor informed all three of them, in an impromptu meeting at Matt's bedside, that he would probably breathe on a vent for the rest of his life.

"He had no right. It's only been a month since I was . . . stabbed," Matt tried to yell through his trach tube. "How dare he? I'll do it. I'll get off the vent."

Ellen and Pat tried consoling him with a special meal of filet mignon, as eating was one of the few physical pleasures left for Matt. He savored it, making it last as long as he could by asking Pat to cut smaller pieces of the tender meat bathed in a lightly seasoned sauce. For dessert—because this was Atlanta—he had peach cobbler.

Everything that Matt had lived for had been slashed away by the knife—he couldn't even breathe on his own. He told Ellen, "I don't want to live. I'm going to pull the plug."

At his parents' next visit, Matt suggested other ways of ending his life. As any mother would, Ellen broke down and cried. Matt looked at his father with a bitter expression, but Pat refused to collapse in the same way.

On another occasion, Pat brought a meal for Matt and learned that he was being difficult, expressing his anger at the nurses and therapists. Collectively having treated more than a hundred patients with spinal cord injuries, the clinicians understood Matt's issues and were willing to bear the brunt of his volatile anger.

"I'm going to pull that plug," Matt told his father. "I'm not going to live anymore."

"OK, go ahead," Pat said.

"What?"

"You want to pull the plug. Go ahead—it's right there."

Matt physically couldn't pull the plug. He looked at Pat and started laughing.

"Mom cried, but you didn't."

"No, I'm not going to cry, even though I feel like it. But you can't keep hurting your mother. You talk about suicide, and she cracks like an egg because she loves you so much."

"I'm sorry, Dad," Matt whispered through his tube.

Long-term ventilator management involves using a tracheostomy tube with a special valve that has a hole in it to allow air to enter from the mouth. Its opening is capped to prevent the voice from turning into a hiss of escaping air. The tube itself does not go into the larynx, unlike a regular trach tube for a ventilator, so it allows the vocal cords to vibrate against each other and produce speech. In theory, for short periods, Matt should have been able to talk with this kind of trach tube. But he couldn't. His vocal cords had been damaged by the trach tube for the ventilator, almost scarred into silence.

Matt was sent via a tunnel to the neighboring Emory hospital, where an ENT specialist used a laser knife to remove the scar tissue. Matt waited to regain his voice. He received medications to reduce the inflammation, and the respiratory therapist watched over him to ensure the best conditions for his recovery. Then it was time to see if the laser surgery had worked, to fit Matt with the special trach tube, and to hear him speak. Matt spoke briefly; but the scar tissue eventually returned, and his speech was limited to a hoarse whisper. To get someone's attention, he had to click his tongue, the frequency rising with the urgency of his needs. He also learned to press his head against a pressure-sensitive pad, which activated a beeping call light.

Another surgery might excise the scars, said the ENT surgeon, so Matt again went under the laser knife. Again, he and his parents and his rehabilitation team waited for words to blossom out of his throat. Again, the buds whispered their refusal.

Early on, Ellen and Pat heard from Matt's clinicians that his injury was complete, usually followed by "I'm so sorry" and often accompanied by a hug. At first, the Nagles hadn't understood these comments. They were just learning about complete versus incomplete spinal injuries and the implications for treatment and recovery. Matt's condition and prognosis were among the worst of all the patients at the center. Ellen saw people wheeling around in manual wheelchairs; they had spinal injuries at the chest level or even low in the neck and had limited arm use, and Ellen thought that they were so lucky. When they saw tetraplegic patients who weren't on ventilators, they would have given anything for Matt to be fully paralyzed—but not on a vent.

Having met the other patients, the Nagles realized that, until now, they had tended to see disabled people almost as if they had been born that way. But they hadn't. A tragic story propelled each wheelchair.

An eighteen-year-old man had fallen off a boat and gotten pinned against the dock. His spine was crushed, causing incomplete tetraplegia and ventilator dependence. Due to their similar circumstances, he and Matt connected right away. He was married and had two children, but there was no family support, no money, nothing. Shepherd Center provided his care at no cost, as part of its charitable mission. This kind and empathetic young couple frequently visited Matt. Feeling sorry for them, the Nagles provided nice dinners and other treats. As this young man's bruised and inflamed spinal cord healed, he got off the ventilator, regained his strength, and eventually walked out of Shepherd Center.

A fifty-year-old man's aortic aneurysm had ruptured when he sneezed, shutting off the blood flow to his spinal cord, and he became paraplegic. A well-known writer, he had a partner who helped with his rehabilitation in spite of severe emphysema that necessitated a walker. Somehow, this couple managed to preserve their sense of humor. They knew Matt was having a hard time, and they made a point to see him every day, tell him jokes, and make him laugh.

Disability was a foreign country for Ellen and Pat, a place to which they had never traveled—until Matt got hurt. The Nagles realized that most people didn't know a thing about tetraplegia, and even some doctors knew little about this strange condition, with complications such as spasticity and autonomic dysreflexia.

In this new country, their son was a desperate citizen. They knew Matt could be stubborn, a feature that was his greatest strength and also his greatest weakness. Sometimes Matt just didn't work with his physical and occupational therapists. On occasion he yelled hoarsely at the doctors and was abusive as he disagreed with their recommendations. Matt struggled with the nurses and doctors about his bowel program and thought they didn't know how to manage his care. His bladder catheterization schedule also led to friction with the staff, because it interfered with his favorite soap opera and late-night TV.

A psychologist explained that Matt's frustration stemmed from having almost no control of his life, except with his schedule and treatments.

"We've seen this many times. We give patients as much control as possible—as long as it doesn't damage their health."

Ellen and Pat thought the clinicians at Shepherd were wonderful. Even when Matt was terrible to deal with, they maintained a dialogue. Having treated many patients with this catastrophic injury, this team knew that Matt's emotional wounds were just as disabling as his physical ones. His unpredictable anger could suddenly be hurled at the most convenient target, and the Shepherd staff had learned to not take it personally.

Once, Matt said, "I hate Atlanta." His comment darted out, unprovoked.

"You're in a hospital room," Pat said. "You don't even know what Atlanta is like."

"They all talk so slow here," Matt said.

Pat laughed. These were the nicest people he'd ever met, who provided skilled rehabilitation and tended unfailingly to Matt, even when he was in a foul mood.

"Come on, Matt," he said. "So they've got accents, and they talk a little slower. You're lucky to be here, and you know it."

"You're right, Dad," Matt responded in a Southern drawl.

As Matt's rehabilitation progressed, the Nagles were able to take him out of his room. Their first excursion was to the center's cafeteria. After they had eaten, Ellen said, "Let's take him outside. Matt really hasn't been out since he got hurt."

"OK," Pat said. "I saw a ramp going down into the garden." He led them to the entrance, which was wheelchair accessible but somewhat narrow.

"Be careful. We're new at this," Ellen said.

"Don't worry," Pat said.

At the top of the ramp, each of them thought the other was holding on to the chair, and somehow Matt got away from them and started to roll down the ramp. But it was too late to chase him down, and the weight of Matt and the wheelchair gave him extra momentum as he sailed away from them and kept rolling.

"Even in a wheelchair you're hard to keep up with!" Pat said to Matt at the bottom of the ramp. They laughed in the face of disability, a turning point in their rehabilitation. Matt was racing with paralysis.

Matt often got lost in depression, but the lights that drew him out shone from many friends and neighbors in the South Shore and Cambridge who were eager to help. They knew that the Nagles were suffering emotionally and faced enormous financial burdens: the expense of home renovations, a motorized wheelchair, an accessible van, private nursing care, and all the incidental costs of catastrophic disability. So they decided to hold a fund-raiser.

Pat's brother, Rick, took the lead in organizing the event at the Weymouth Elks, the largest local venue. Supporters and donors also came from Pat's colleagues in Cambridge, the navy police, and other law-enforcement groups.

Pat casually asked his brother, "Who's going to be there?" He hoped there would be at least a hundred people, not because of the money but for the sake of Matt's spirits.

"Well, we're working on that," Rick replied.

"Whatever happens, we're very grateful."

"You guys just show up. We'll take care of the rest."

Early in September, the Nagles flew up from Atlanta to attend the fund-raiser, about which they knew very little other than its location. At the Elks lodge, they had to park three blocks away because there were so many cars.

"Something big must be going on in another hall," Pat said.

"I hope it doesn't affect Matt's event," Ellen said.

When Rick and his committee of organizers greeted them, Pat and Ellen realized that all the cars were there for Matt. People were packed in from wall to wall, barely able to move around. Guests overflowed into the hallways and outside onto the lawns. Pat's buddies from Willow Street in Cambridge, where he had grown up in a two-family home, came to show their support. The disc jockey turned out to be Manny Ribiero, a childhood friend Pat hadn't seen in years. Also in attendance was Casey Ross, a journalist who had written a series of moving stories about Matt for the *Patriot Ledger*. Matt's nurses from Boston Medical Center and his high school friends were all there. And although the Boston Red Sox and the New England Patriots couldn't attend, they sent letters to cheer Matt up; so in spirit, they were there, too. More than six hundred people came for "Miracles for Matthew," and they enjoyed the delicious food catered by the Elks staff.

The guests reminisced about his life before the stabbing. One of Matt's oldest friends, Jared Gray, remembered a snowfall in the winter of 1992, when the flakes came down thick and heavy, forming a foot-high layer of snow that caused school to be canceled.

"Jared, let's go sledding," Matt had said as he stared out at the large, soft snowflakes that were piling up. "What about the hill at the Pine Street cemetery?"

They raced each other to the cemetery, neck and neck all the way, then stopped to catch their breath before climbing up the hill. The snow was like frosting on an angel cake baked just for them. At the top, they found a spot with a clear path, looked around for oncoming cars, pushed off with their hands, and slid onto the street; then they scurried back to the safety of the snow-covered sidewalk. Their plastic saucers had ropes in front, giving the illusion of control as they went spinning down. Sometimes the steep hill shot them down so fast that they rolled off the saucers, but they were used to cuts and bruises.

They must have repeated the run at least twenty times, and after an hour their clothes were wet with snow. Luckily, the only car on the road had seen them near the bottom of the hill and slowed down, honking a rebuke. Jared was cold, breathing heavily, and starting to get a little tired.

"Just one more run," he said to Matt.

Also worn out, Matt dusted the snow off his pants. "OK, one more shot."

So they again climbed the hill on Pine Street, and Jared yelled, "Go!"

Matt pushed off hard and careened ahead of Jared, certain to win this last race, flying down the hill, the heavy snowflakes all around him—until he rolled off the saucer into a pine trunk.

Matt felt a sharp pain in his thigh. He stood up—the pain got worse. Massaging the limb, he realized that his jeans were torn and his thigh felt wet and sticky.

"Hey, Jared, come look!" he yelled out. Matt knew he was in trouble.

Jared came running over. "Matt, that's gross. You're bleeding."

Matt's pain was excruciating. The boys realized that they should get home as soon as possible, but Matt could barely walk.

"Stay here. I'll run and get Mike." Jared made an ice pack out of the soft snow and applied it to Matt's thigh. "Maybe it'll help the bleeding."

Matt waited on the hill for about ten minutes. The snow was coming down more heavily, and it was getting dark. Soon Mike and Jared drove up slowly in the Mustang and helped him into the backseat, where he lay in pain until they got to South Shore Hospital. The Mustang didn't handle well in the snow, and every skid on the roads was a stab in Matt's thigh. How was he going to tell his parents? What was going to happen to his leg?

At the hospital, Matt was rushed through the large glass doors. The ER nurse in blue scrubs took one look at Matt, helped him onto a stretcher, prepared the sterile suture kit on a stainless-steel table, and called the doctor. By this time, Ellen had arrived from her job in Boston. She cried as the nurse brought her into the ER.

"That tree trunk made a four-inch gash in your thigh," the white-haired doctor said to Matt. He cut away Matt's pants and squirted a clear fluid from a large syringe to clean off the wound. "I'll numb the area before I put in the stitches, so you'll feel a couple of needle sticks."

"OK," Matt said.

The kind nurse had given him some pain pills, and soft snowflakes fell in his head as he began to feel sleepy. Matt barely noticed when the doctor used a needle to check if he felt any pain; then he trimmed away a jagged edge of Matt's wound with his scalpel.

When it was all over, the doctor said, "You still need a tetanus shot, and here are some antibiotics. That's going to hurt for a while, so I'll give you some pain pills. And no more sledding for a while."

"Thanks," Matt said. "But what about hockey? I've got a big game in two weeks."

There were many other Matt stories, as his friends tried to convince the Nagles that Matt could recover from just about anything. And the fund-raising went on through the night. The DJ sold hundred-dollar tickets to auction off prizes donated by local businesses. Between the cover charge and the auction, the event raised almost fifty thousand dollars for Matt's medical expenses. A case manager at Shepherd had told the Nagles that the lifetime-care costs of people with tetraplegia were $3 million. Rick sent a video of the fund-raiser to Matt so he would know how many people were pulling for him.

"Everyone's been so good and generous," Ellen said. She recalled the movie they had watched in Atlanta, and she said, "We've got to pay it forward."

Contributions to Matt's fund came from some unexpected sources, as Pat accidentally learned. (Like any top-notch detective, he did not reveal his informant.) A fellow policeman said that when his visiting brother-in-law wanted to go to a strip club, they had gone to Brockton, to be incognito. On the stage, some young women were dancing near a hat and a sign that said DONATIONS FOR MATTHEW NAGLE.

Some donations in kind were tremendously meaningful. Matt's friend Kara gave them four tickets to a Red Sox game. They were great seats, the best incentive for Matt to finish his rehabilitation and return home.

First, the Nagles had to renovate their home in Weymouth, to make it wheelchair accessible and comfortable for Matt. But they had to stay in Atlanta to support Matt, and getting a construction company to do the work would be prohibitively expensive, even with the fifty thousand dollars from the fund-raiser.

On the phone with their son Mike, the Nagles discussed how they could get the home ready for Matt. He couldn't stay at Shepherd for more than two months, and they refused to send him to a nursing home.

"I'll do it," Mike said. "I'll draw up the plans and clear the land for the driveway and the addition."

"Mike, it's too big a project," Pat said. "You've done a lot of construction work, but this is way too much for you."

"Dad, I can do it. I'll get help. It'll be done by the time Matt gets home. Contractors can be slow."

"OK, Mike, but get some help," Pat said.

A few days later, Mike called back with his preliminary plans. He wanted to fix up the lower level with windows, wallpaper, and a bathroom with tiled floors. Adding a wheelchair-accessible entrance and another driveway would mean having to take down an oak and some maples. For Matt and his friends to hang out and have a few beers, Mike designed a deck that would lead from the addition to the house.

Pat returned home to help with the renovations. He stood at the far side of the house, near the three-season room where Mike had laid out a driveway, wondering how the trees would be cut and hauled away. Paul, a neighbor, came over to chat. Pat explained his dilemma and said that he would have to hire some people for the trees.

"Look, I want to do something for Matt." Paul stared up at the trees blocking the sun. "You get the bucket and wood chipper. Mike and I will do the rest."

"But that's more than a dozen trees," Pat said.

"Don't worry. We'll take care of it."

A week later, Mike told his parents that he and Paul had done the project in a day. It was time to grade the driveway.

By the end of September, Mike had finished the lower level, with its accessible entrance and a roll-in shower for Matt. A new driveway led up to the entrance for Matt's new apartment. Any person in a wheelchair would have been thrilled with this dwelling, but Mike was not finished. He wanted to add on a room with an entrance through the kitchen and an exit onto a deck. Pat couldn't believe what his son had done, just in time for Matt's return home.

"But Dad, there's still the dining room."

"Mike, you've done more than anyone expected. When Matt's home, we'll worry about that room. Maybe it can double as a living area with a TV."

<hr />

Ellen and Pat had a meeting with Matt's psychologist at Shepherd Center shortly before he went back home.

"Are there common denominators among these patients?" Ellen said.

"Yes. At this stage, most patients with spinal cord injuries are depressed," the psychologist replied. "Everyone has unique responses, but there are certain threads that go through the early stages."

"So much anger," Pat said. He shook his head. Ellen sighed.

"Yes," the psychologist said. "And the worst part is the unpredictability."

"How long does it last?" Ellen asked, hoping that this phase of Matt's adjustment would resolve soon.

"Good question," the psychologist said. "I wish there was a definite answer, but there are so many factors. Some recover soon; others remain trapped in their anger."

"Are there different stages that people with spinal cord injuries go through?" Pat said.

"It's like the process of grief," the psychologist said. "There's denial, then bargaining, then anger. But that can vary. If patients are lucky, someday they accept the condition of being quadriplegic or paraplegic."

"How long does that take?" Ellen expected to hear anywhere between three and six months.

The psychologist looked down at his notes and said quietly, "Five or six years."

"Five or six years?" Ellen and Pat echoed.

"If people can get beyond that five-year mark, they generally do well," he explained. "They seem to go back to their level of happiness before the injury. They can plan for the future. It's time to look ahead and make the most of their lives. But some people just don't make it that far because of complications."

"We'll watch out for them," Ellen said. "The staff has taught us well."

"I have to warn you about what you're planning," Matt's psychologist said as they ended the meeting, his expression serious and his voice somber.

Ellen and Pat looked at each other, their eyebrows slightly raised.

"I don't know how stable your marriage is—and I don't want to pry—but the stress of caring for someone on a vent is incredible. The divorce rate is very high."

"We've talked about it. We have a lot of support," Ellen said, and she took Pat's hand.

"We'll be the exception," Pat said. "We can do it. Our son's built an addition to the house for Matt. Our friends and family had this huge fundraiser. If you'd seen the event they put together for Matt, you'd know what we mean."

Ellen smiled at Pat, then looked over at the psychologist.

"Thanks for the warning, but I think we'll be fine."

Parade with the Patriots

Matt, Late 2001 to Early 2002

Although Matt's severed spinal cord still crackled with electrical activity like a downed power line, he regained no movement, not even a muscle-fiber twitch. He could breathe for only two minutes off the ventilator. At Shepherd Center, his dedicated nurses, therapists, and doctors could not reverse the damage of the knifepoint embedded in his spinal cord. When the Nagles felt ready to care for him, Matt went to a transitional facility before going home. On October 1, 2001, three months after being stabbed, he was discharged to a nursing home in Massachusetts while his parents searched for at-home health care.

At this facility, a dozen miles from his home, Matt learned about the nursing shortage firsthand. He fell asleep early one night and woke up twelve hours later to discover that he hadn't been turned. Because he couldn't feel pain or move, Matt developed a pressure sore over his tailbone. Upon confronting his caretakers on the night shift, he got an "Oh, well" response. The skin ulcer took two months to heal.

Matt's room at the nursing home was filled with flowers, gifts, and cards. Meaghan Murphy, Katie Perette, and other friends came in waves. Even the Nagles were surprised at their son's popularity. Katie made a poster for Matt, a collage of photos with his friends, to hang in front of his bed. But, at her later visits, there were not as many visitors, and the waves slowed down to ripples as the months passed.

On weekends, the nursing home had temporary help in the form of caretakers who were well meaning but not especially well trained. One Saturday, the Nagles walked past a room, heard a noise, and saw that an elderly woman had been dropped out of a lift. Later they discovered that the facility had changed its name, apparently trying to shed its bad reputation but having limited success. Eager to free Matt from this place, the Nagles searched every avenue and called many contacts, but at-home nursing help was impossible to find on short notice.

Matt's breathing was their biggest worry. His VIASYS TBird ventilator sent oxygen in a blue tube to his throat, and a white tube sent carbon dioxide from Matt into the machine. If its battery died outside the nursing home—as it did on rare occasions—Ellen and Pat took turns giving Matt oxygen from a green metal cylinder, squeezing the rubber bag attached to his tracheostomy tube twelve times per minute until he reached the lifeline of electricity to recharge the machine.

When Pat stopped in after work one night, Matt was having trouble breathing, even on the ventilator. Pat called for a nurse, but no one came.

Matt gasped and whispered, "I'm dying. . . . I'm going to die."

A nurse finally came to suction his lung secretions but didn't seem too concerned. Although they never discussed it with Matt, his parents were horrified at the possibility of a ventilator malfunction—a terrible way to die, choking and gasping for breath. For Matt's safety, Ellen started sleeping on a floor mattress at the facility. When a respiratory therapist told her to watch out for the cockroaches, she said, "I'm out of here, and he's out of here." Pat agreed.

The Nagles brought Matt home. Since this went against the advice of the facility's clinicians, they had to sign a waiver and accept all responsibility for his care. Mike had labored night and day to finish the home modifications for his paralyzed brother, but Matt had not yet seen the renovations. The wheelchair van drove onto a new, smoothly black-topped driveway. In the autumn sunlight, Matt missed the shade of the oak and the maples and realized that he might never again go wading in the backyard pool of red, yellow, and orange leaves.

Mike wheeled Matt into his new apartment through an entrance that had been widened for his motorized chair. Next to his bed was a Hoyer lift, a sturdy, stainless-steel base with a hammock to transfer Matt from the bed into the chair. The new shower stall was designed for a wheeled commode that also functioned as a shower chair. On the walls were framed sports photographs of Matt, including the one of him chasing down the running back with the impossible head start. Best of all, Matt had a widescreen TV for watching ESPN and NESN, a New England sports channel that showed the Red Sox games.

"Mike, I don't know what to say. Come here; I'll give you a hug." Then Matt realized that he would never be able to actually hug anyone because his arms were paralyzed. Well, he could shrug his shoulders toward them to show his affection.

Mike went to Matt. As he did so, Ellen came over to her boys for a group hug.

"Mike, thanks for this wonderful gift," she said. "I'm so proud of you."

"Here's the ramp to the backyard," Mike said, guiding the wheelchair outside.

Matt stared up at the pine and oak trees, at the hedges that protected the backyard, at the sun umbrella that would give him shade and prevent overheating. He had learned that his spinal cord injury had cut off his brain's thermoregulation and made him poikilothermic; his temperature, like a frog's, would be the same as that of his environment.

"Mike, it's too good to be true. How'd you do all this?"

His brother only smiled and patted Matt on the shoulder.

At that moment, Matt realized how much his brother loved him. They had been through plenty of brotherly fights, and he usually lost because Mike outweighed him. As they got older, Mike had disapproved of his brother's wild ways and lack of direction. Matt knew that he was just being an older brother, trying to help him grow up, but he had resented Mike for being stronger and more sensible. Mike had never said that he loved him, but his feelings for Matt were articulated in the wheelchair-accessible bathroom, the ramp, and the widescreen TV.

The deck was perfect for hanging out with friends; they'd have a few beers, talk about old times, discuss their plans for the future. Then it hit Matt like the rush of a linebacker. What plans did he have? He had been sent to the electric chair—for life. But he recovered and tried to smile at the inscrutable face of his future.

<hr>

The first night after fleeing the nursing home, Pat and Ellen were getting ready for bed, when Matt said, "Dad! Dad!" His tongue clicked rapidly, his sign for urgent help.

"What's up, buddy?"

"Not breathing!" Matt forced out the words through the plastic tube in his throat.

Pat looked closely at the machine, which hissed and puffed as usual, and he assured Matt that it was working.

"Dad, can't breathe, can't," Matt insisted in a choked, sandpaper voice.

"Don't get excited." Pat saw that the machine was working, but for some reason the oxygen wasn't getting to Matt.

"I'll bag him," Ellen said. She connected Matt to the oxygen tank and began forcing the air into his lungs. "Call 911! It's an emergency!"

"No ambulance." Matt shook his head. "No hospital—they'll keep me. I'll get pneumonia, a bedsore."

Ellen wasn't going to send Matt back to a hospital. She handed Pat the gray rubber bag and called a respiratory therapist who had worked with Matt. He dropped everything and drove up from Plymouth, but for an hour the Nagles alternated breathing for their son to keep him alive.

It turned out that the settings on the ventilator had malfunctioned. If Matt had taken sleep medication, the respiratory therapist said, he would never have woken up. For some reason, the vent alarm didn't go off.

"They did something different," Pat said, referring to the staff at the nursing home. "What did they do? They didn't tell us. They changed something but didn't tell us."

Ellen hugged her husband. It had been a close call.

They still had no nursing help, so Ellen took care of Matt, sleeping beside him in the basement on a cot. Sometimes, in the middle of the night, she heard the ventilator alarm and jumped out of bed to check on Matt, to watch his chest move as air was pumped in and out. But even when Matt was safe, she stayed awake, unable to go back to sleep, the adrenaline coursing through her blood. Ellen took care of Matt for his first two weeks at home, relying on the skills she learned at Shepherd Center, referring to her notebook, running on pure love for her son.

For him, it was a difficult transition. Matt was angry about everything: his paralysis, the tube forced into his throat, the loss of control over his bowels and his bladder, his bedsores, his lost sexual function, his depression, his bed cage. His anger swooped down on his parents, the most visible and most forgiving targets. Once, he told Pat that Ellen had twisted his arm out of its socket, a completely irrational comment in light of all that his mother was doing for him. That day, Ellen had been completely exhausted. She went downstairs, adjusted his pillow for at least ten minutes, and thought to herself, Please, God, I'm trying, but I can't do anymore. No more, no more.

Matt's needs were nonstop, and if they weren't met quickly, he lashed out.

||||||||||||||||||||||||||||||||||||

Some of Matt's friends continued to visit on a regular basis, even after the final dramatic article in the *Patriot Ledger* and the excitement of his homecoming. Meaghan Murphy was one of those faithful friends, and she often came to help the Nagles. The first time Meaghan fed Matt a meal, she held a glass with a straw while he drank his Gatorade. Then she prepared his favorite meal: lobster. The experience was awkward, as Meaghan had to deal with the mechanics of cutting up the food and placing it in Matt's mouth. At first, she was so nervous that she cut the pieces too small.

"Meggie, you can cut them bigger," Matt said. He smiled and shook his head. "They pulled out the feeding tube for a reason."

When she dipped only a small part of the next piece into the melted butter, his smile widened. "Meaghan, just put the whole thing in there."

Meaghan laughed at herself, feeling a little embarrassed about her awkwardness. But once she got into the rhythm of feeding Matt, it brought them closer, and she felt happy because Matt seemed to enjoy the lobster even more because of her presence.

||||||||||||||||||||||||||||||||||||

When the Nagles finally found some good nurses and physical therapists, Matt discovered new targets for his anger. He attacked them verbally and sometimes even spat at them. Raging inwardly at the cowardly stabber, Matt kept his anger simmering. He could never tell when it would boil over and scald the people closest to him.

Eventually the Nagles found a nurse who understood Matt, a kind and patient young woman named Gina. Bubbly and upbeat, she also helped out when friends came to visit Matt, which allowed Ellen and Pat some respite and a chance to enjoy themselves. Matt's most serious girlfriend had already signaled that she wouldn't continue the relationship, and Gina filled a void as a young woman who cared deeply for him.

Matt's initial publicity ebbed as other events caught the attention of New England reporters. Pat desperately tried to keep his son's name alive. He wanted him to be remembered and considered in case new research might bring even a flicker of hope for recovery. Pat and Ellen didn't want him to be just another victim, an atrophied life.

At first, Matt was angry with his parents about all the media attention. But he settled down when Pat said, "Matt, if we stop putting your name out there, you're basically pushed into a corner."

Enjoying what he could in the existence he had left, Matt often listened to music, as if aural stimulation could help make up for the loss of all sensation below his shoulders. One of his favorite musicians was DMX, who combined energetic beats, spoken raps, and bluesy melodies; his lyrics mixed the sacred with street life. And Matt still enjoyed the classic rock of Led Zeppelin and Aerosmith.

After attending a Britney Spears concert in December 2001, Matt was thrilled to meet the singer, who kindly sent him a signed photograph that he proudly displayed in his room (see figure 5.1). On occasion, people who did not recognize Spears asked if the photo was of his girlfriend. Matt smiled and winked.

During this time, Matt's team at Shepherd Center stayed in touch by e-mail. With the enthusiasm so typical of people who treat severe disabilities, physical therapist Diane Johnston yelled out into cyberspace:

HEY MATTTTTTIIIIIIIIIIEEEEEEE!!!!!!!!!!!!!!!!!!!!!!!! How are you doing, man???? I miss your smile around here. Drop me a line and let me know how things are going for you up in Boston.

Cynthia Skinner, Matt's occupational therapist, inquired after her patient and wrote:

I hope you are SITTING UP!!!!!!!!!!!!!!!!!! :-) Love you guys.

Ellen sent an update and asked the therapists to let her know of any new, cutting-edge technology that might help Matt's condition. At Shepherd Center, some clinicians had described a lung pacemaker to her. In December 2001, she asked William Dobelle, a pioneering researcher with respiratory devices, if the phrenic nerve pacemaker could replace the ventilator and if anyone in Boston had experience with this device. No one did, but Ellen continued her research. The Nagles were willing to travel to New York for the surgery.

In addition to his lungs, Matt had to contend with one of the worst losses for young people with a severe disability—that of his sexual function. His severed spinal cord allowed Matt to only have reflexive erections, and he experienced no sensation. Attempts at intercourse could lead to autonomic dysreflexia, with a pounding headache caused by extreme blood pressures and the risk of a stroke from bleeding in the brain. (Some people are luckier, in a relative sense. One of my patients with an incomplete spinal injury said that he could have intercourse and that it was a very satisfying experience. Most important to him was not the act itself but the closeness with his partner.)

A few of Matt's friends continued their emotional support. Kara had sent Matt many encouraging e-mails while he was at Shepherd Center. Her father also had a spinal cord injury, so she understood Matt's condition far better than his other friends. After Matt returned home, she stayed close to him. Once, Ellen found her lying next to Matt in bed, and she wondered if in fact they also had a physical relationship. But he always said that they were just friends.

Matt once told Pat about another tetraplegic person, with an incomplete injury and the ability to use a joystick for his wheelchair, who met a woman with a spinal injury on the Internet. Their personal-care attendants arranged a meeting, and they were able to enjoy making love.

"You know, it can happen," Matt said.

But the Nagles weren't particularly keen about arranging a similar tryst, even though Ellen would have loved a grandchild. Some of the young women the Nagles met after Matt's injury would say they were his girlfriends. For Pat, it became a family joke: How had his son juggled his social calendar, and how was he still doing it? And without a BlackBerry!

Even though they joked about Matt's social life, the Nagles realized that friends and relatives were vital to his well-being. Ellen's sister Lee Ann was visiting from Connecticut and would be good company to ring in the New Year.

On New Year's Eve, the City of Cambridge needed extra help with security, and Pat volunteered because the medical expenses were piling up and the money was running out. Matt wanted to have some friends over, and his parents agreed.

For his brother's privacy, Mike had built a side entrance to the basement apartment, so Ellen had no idea how many people had gathered there. As the party got into full swing, the music started. Ellen was glad they were having a good time. Maybe that was why Matt had said, "Mom, make sure that door stays closed." She had no problem with the noise—after all, it was New Year's Eve.

When the music became louder, Lee Ann opened the door to the basement and, out of concern for the neighbors, asked Matt to turn down the volume. In the process, she saw something, quickly shut the door, and started laughing. She didn't say anything to Ellen.

When Pat returned from his overtime detail and the party had ended, Lee Ann asked if Matt often had parties downstairs. Pat said that he didn't.

"It's nice of you to let him bring in strippers," she said casually.

Pat went off to confront Matt, who said, "What? Are you mad 'cause I didn't invite you?"

That was the last such party at the Nagle house.

When Matt saw a segment on the news about stem cells, he yelled for Ellen as loudly as his tube allowed. Researchers at Advanced Cell Technology (ACT) in Massachusetts used nuclear transfer to achieve human therapeutic cloning. Eggs were collected from women's ovaries, as with in vitro fertilization, and their nuclei—which contained the genetic DNA code—were removed. Researchers at this corporation then transferred nuclei from adult donor cells into the empty eggs. Chemical stimulation started cell division, the process that normally occurs with a fertilized egg. An egg divided until it consisted of six cells, proving that even mature human nuclei could be reprogrammed to an embryonic state.

Michael West, the president of ACT, hoped to create stem cells that could be transformed into a variety of specialized cells to treat spinal cord injuries, Parkinson's disease, and other conditions.[1] These cloned cells would be genetically identical to the donor, which would reduce the risk of immune rejection. In contrast, of the almost sixty human embryonic stem cell lines available for research supported by federal funds, all would probably be rejected by patients' immune systems.

Stem cells are the fundamental cells of life that can replicate themselves in a special culture medium or transform into specialized tissues. At present, there are five sources of stem cells: embryos, newborns, adults, therapeutic clones, and genetically engineered stem cells.

Like all animals, humans start off as stem cells. Five days after a sperm fertilizes an egg, the embryo of a hundred cells is smaller than the period at the end of this sentence. About a third of these are stem cells that can form any of the body's tissues; however, they cannot treat a damaged organ because their primitive state could also lead to tumors. For that reason, embryonic stem cells have to be carefully selected for therapeutic purposes.

Newborn stem cells are derived from the umbilical cord and placenta, but in small numbers, which allows the treatment of children with blood cancers but limits their potential for treating adults. Biotechnology companies are working on techniques to increase the yield of stem cells from this source.

As adults, our bodies have stem cells that regenerate tissues; these have been found in the bone marrow, intestines, lungs, liver, brain, and spinal cord (see figure 5.2). Our skin maintains its integrity through stem cells that heal cuts and bruises. But we cannot repair our bodies in the way a starfish regenerates a ray. A human amputation is final, and the only option is an artificial limb. That is the price we pay for reaching the pinnacle of the evolutionary tree, with more than two hundred specialized tissues.

The Nagles hadn't even thought about human cloning until Matt's paralysis. The *Patriot Ledger* heard of their interest and ran a story titled

"Cloning Offers Hope to Some, Despair to Others: Paralyzed Wey-mouth Man Sees Prospects for Cure." Matt's parents said they hoped that "research into the cloning of human embryos might lead to developments that will give him the use of his arms and legs again." These findings had "touched off a new debate and not everyone is thrilled with the development." President Bush said, "The use of embryos to clone is wrong. We should not, as a society, grow life to destroy it, and that's exactly what is taking place."[2]

In December 2001, Ellen sent an e-mail to Hans Keirstead at the Reeve-Irvine Research Center of the University of California, Irvine. A tetraplegic woman in Massachusetts named Deborah Murphy had told Ellen about his promising research with stem cells developed by Geron Corporation. They had been obtained from an in vitro fertilization clinic and were destined for destruction.

Ellen asked if "there is hope for my son to recover use of his body within the next five to ten years." She explained, "My son truly believes that it will happen because of people like you, who are dedicated to this mission." Ellen wanted to help raise funds for the research and was interested in clinical trials.

In his response the same day, Keirstead wrote:

I was so sorry to hear of your injury, Matt. You are the reason that I am doing what I am doing, along with the discoveries that give me certainty and encouragement along the way.

Please know that there is more than hope on the horizon. There is almost a certainty that effective treatments will be available, but we don't know when. Near-certainty, because we already have treatments that work on rats and mice. We don't know when, because we have to refine the treatments in some cases to make them better or without side effects, build the human counterparts of the rat/mouse treatments, test those humanized treatments on animals, then deal with the government to attain approvals for working on humans. All those things are doable, and in some cases, well under way. We have hit roadblocks in the past, and I am sure that we will in the future, but you should know that we are very stubborn!

I greatly admire your efforts to help us get there quicker. . . . That old stereotype that scientists work too much and don't get out enough is true, so thank you for knocking on our door and encouraging us.

<hr>

Matt and his buddies from high school watched the Super Bowl at the Nagle house on February 3, 2002. Adam Vinatieri made a forty-eight-yard field goal to help the New England Patriots defeat the favored St. Louis Rams (20–17), one of the biggest upsets in football history. Tom Brady completed sixteen of twenty-seven passes for 145 yards, including a touchdown, and was named MVP.

After the game, Matt said, "Dad, there's going to be a party at city hall for the Patriots."

"I know," Pat said, "there'll be about four hundred thousand people."

"I want to go."

"Absolutely not. No way are you going."

"I'm going," Matt insisted.

"Well, you can try, but you won't get near there."

Pat was on duty when he got a call from a Boston police officer. "Are you Sergeant Nagle, and do you have a disabled son?"

"Yes, Matt's in a wheelchair," Pat replied. "Why do you ask?"

Matt, his nurse, and his friend Rich Rosado had driven to the parade in a wheelchair van. When the officer told them to leave, Matt called him over and said, "My dad's in charge of homicide in Cambridge—Sergeant Nagle."

"So?"

"I want to see the Patriots," Matt had told him.

Chatting as one police officer to another, Pat gave him the lowdown on the paralyzed former football star and admitted, "Yes, that sounds like Matt."

"OK, we'll take care of him," the Boston officer said.

The officer brought Matt right into the center of the celebration, where two Patriots cheerleaders from Weymouth came off the floats to give him hugs and kisses.

Bob Kraft, the owner of the Patriots, came over and said, "You can have any one of the Patriots come here to see you. Who do you want?"

Matt looked at the girls and joked, "I know who you want: Tom Brady."

So the quarterback came to visit Matt. Then Bob Kraft brought over the Super Bowl trophy and held it close to Matt, who gave it a kiss.

Pat heard about all this later and realized that, while he was worrying about his son, Matt was having the time of his life.

Matt formed a special connection with the Patriots at that event. When Matt was feeling especially discouraged, John Dennis, a sports announcer on the radio station WEEI, mentioned Matt's struggle to a friend who was on the team's staff. As a Christmas present, Matt got tickets for a Patriots game. The radio announcer also organized a video in which many of the Patriots cheered him up. Tedy Bruschi gave his famous smile and said, "Hi, Matt. I know you're a little bit down, but hang in there." Quarterback Tom Brady told Matt, "Stay strong. We're all pulling for you!" The news crew for a local television station, Channel 5, came to film Matt for the nightly news after hearing of the Patriots' video message. And Chris Berman, the host of *Monday Night Football*, encouraged Matt with his famous line, "You could go all the way."

                                               IIIIIIIIIIIIIIIIIIIIIIIIIIIIIIIIIIIII

For Matt, paralyzed and cut off from the sports he loved, gambling was the next-best thing. Occasionally he argued with his father over money. Pat knew he wanted it for sports betting, but the family had enormous medical expenses.

Over time, their discussions became more heated. One day, Pat drew the line. He offered money for whatever Matt needed, except gambling.

"No, you know what?" Matt said. "I hope you die. I hope you die of a heart attack."

"You'll probably get your wish at some point." Pat had just received a cardiac stent to keep an artery open.

"I want you out of here. Get out." Matt muttered something else as his father left.

"Did you hear what he said?" Gina asked Pat.

"I don't even want to know." Pat quietly left the room.

Matt dreamed of getting off the ventilator, of escaping its hissing reminder of his paralysis. In the spring of 2002, he had exploratory surgery to see if he would benefit from a pacemaker for his phrenic nerve, which would rhythmically stimulate his diaphragm and perhaps allow him to spend at least a few hours off the vent. But the surgeon could not even find the nerve because of scar tissue left by the attacker's knife.

Because his voice was still damaged by the tube in his throat, Matt had vocal-cord surgery, and a stent was placed to help the tissues heal properly after the scar tissue was cut away.

Ken Aron from Avery Biomedical Devices, the company that manufactured the pacemaker, inquired about Matt's surgery. Ellen replied via e-mail: "Matt is waiting to have the stent removed. I haven't spoken to him regarding the phrenic nerve because he's been very depressed of late. I'll let you know as soon as I know."

Matt's bitterness corroded his relationships with caregivers, the constant reminders of his helplessness. Pat tried to help Matt understand that these health care professionals were his friends and he couldn't afford to lose them. Some people he had thought were his friends were no longer there, but his caregivers and his family remained.

"If you don't treat them right, you'll end up in a facility," Pat said.

"I don't care." Matt turned his head toward the window. "Do what you want."

Many personal care attendants (PCAs) came to help the Nagles take care of Matt: feed him, bathe him, do his bowel care, clothe him, and transfer him to the wheelchair. And many left. But they all tried their best to help the angry and sad young man. All were caring people, capable of handling his basic bodily needs.

If an attendant called in sick, Ellen slept downstairs. Vigilant about listening for the ventilator alarm, she never reached the stage of deep and restful sleep while monitoring Matt. She got up every two hours to turn Matt so he wouldn't get a bedsore.

Life had changed for the Nagles in many other ways. Pat realized that he had to do something about his two Rottweilers. He loved his dogs and had spent a lot of time training them, but with nurses and attendants coming and going, the dogs often escaped through the front door. Pat gave them away, along with their kennels and supplies.

In spite of the emotional firestorms and the chills of depression, at least the Nagles had Matt in their own home, and they had happy moments together. Mike continued working on the house, and his project concluded with the addition of a living and dining room. From the kitchen, the ceramic tile floor in the hallway seamlessly changed to a floor made of Brazilian cherry to allow easy wheelchair access. A ceiling fan with light fixtures helped circulate the air, both for cooling and heating. For the winter, Mike had added electric baseboard heating, but there was also a fireplace with an ivory stone mantel. Because Matt's internal mechanism for controlling body temperature had been stolen by the knife, Mike wanted to ensure a comfortable environment. French doors led out over a ramp onto a wooden deck, where Matt could hang out with his friends. Mike had kept his family out of the construction area until he was finished, so it was as if they had entered a new house.

"Mike, we wanted a living and dining area. But this is the Taj Mahal," Pat said.

"Well, Bob and Carol Barry donated the tiles and the wood," Mike said with his typical modesty.

Matt was speechless, for once, but his smile said everything.

||||||||||||||||||||||||||||||||||||||||||||

Because of his paralysis, Matt had to depend on others for a variety of activities. One day he asked his father to pick up a prescription at the CVS in Weymouth. Pat offered to go to the smaller store in Hingham, where he didn't expect a line of customers, but Matt said the prescription had already been called in.

Pat strolled into the drugstore, which was packed with shoppers, some of whom he knew. He asked for the Nagle prescription and then wandered away from the pharmacy counter to browse. When he heard the pharmacist announce his name, he rushed back.

The pharmacist handed him a bag and said, "Your prescription's ready."

Pat looked inside. "No, this can't be right. It's Viagra. This prescription's for my son. He's quadriplegic."

"Is your son Matthew Nagle?"

Pat nodded, too surprised to say anything.

"Well, it's your Viagra."

By this time, Pat's face had turned red, and he glanced around with an awkward smile, hoping none of the people knew him. Luckily, all of the nearby faces, some of them smirking, were unfamiliar.

This kid's going to kill me yet, Pat thought to himself as he briskly walked out.

"Do you know what you sent me for?" Pat said to Matt when he got home.

"Why?" Matt laughed. "Did they say something?"

⸻

After the first huge fund-raiser for Matt, other events took place. Many people in Weymouth and neighboring Hingham overwhelmed the Nagles with their generosity, including neighbors Jerry and Pat DiLillo and the DiLillos' friends Ron and Leslie Pollara. As distributors for liquor stores, Jerry and Ron spread the word about Matt, and donations poured in from their customers, including Martignetti Liquor and Curtis Liquors. Two large fund-raisers were held at Tosca, a well-known restaurant in Hingham that donated to Matt's fund. Jimmy McSweeney, a former cop who was now an insurance agent, organized two successful golf tournaments. The money was placed in a special fund for Matt and used to improve his quality of life in a variety of ways. For instance, when the Nagles had to spend six thousand dollars to repair the lift in their already expensive wheelchair van, the money from the fund helped.

On the anniversary of Matt's injury, the Nagles had a party in his honor, an occasion for his friends to come and visit, drink a few beers, and chat about the good old days. On the deck that Mike had built for him, Matt relaxed with his buddies, enjoying the fresh pine air instead of the antiseptic smell of his perineal cleanser and the deodorant spray in the basement. It was about a hundred degrees outside, and Matt kept saying that he was too hot. For someone who was poikilothermic, the heat could be dangerous—so Pat bought an air conditioner that day.

At the party, a friend recounted one summer afternoon that Matt and his friends spent lounging at Tommy O'Neill's pool. They were soaking in the sun when Hemma stopped by.

"Guys, I'm broke," Hemma said. "I need a beer and a smoke."

Matt winked. "Sure, Hemma, but first, you've got to do us a small favor." Matt whispered something to Rich and Tommy, who disappeared inside.

"Hey, don't make a big deal out of a can of beer and a cigarette," Hemma said.

Tommy returned with a six-pack and a floral pink object. "I think it'll fit."

Hemma looked from Matt to Rich to Tommy.

"Here's the deal," Matt said, grinning. "We'll give you a six-pack *and* a pack of smokes if you wear Mrs. O'Neill's bikini and dive into the pool."

"You're nuts!" Hemma shook his head. "Effing crazy if you think I'll put on that . . . that thing."

"We're 100 percent serious," Rich said with a deadpan expression.

"Yeah, and you've got to let us take a photo," Tommy said. He offered the swimsuit to Hemma, who looked at him in disbelief.

"Hemma, take it or leave it. I'm feeling thirsty, so if you don't want it. . . ." Matt reached for the six-pack lying on the poolside table.

Hemma grabbed the bathing suit from Tommy and went inside to change. He returned wearing the ill-fitting pink bikini and strutted around as though on a runway. The other boys cheered and whistled.

"You fill that out real nice," Matt said, "but you've still got to dive in for a wet swimsuit photo for the *Patriot Ledger*."

"They'd never print this." Hemma paused. "Would they?"

"Never can tell," Rich said. "If I was editor, I'd print it. Local color, you know."

While the other boys laughed, Hemma threw up his hands, went to the diving board, and plunged in. At the other end of the pool, Matt and

Rich were waiting with a camera, a pack of Marlboros, and a six-pack of Coors Light.

Matt's friends had always appreciated his sense of humor, and the pool story brought back memories of other escapades.

By late evening, everyone had left the party, except for Katie and a friend named Stephanie, who rested in bed with Matt for company and comfort.

"I can't believe this happened to me," he said.

All three of them started crying.

"Yes, it's terrible," Katie replied, "but Matt, you're going to pull through."

Matt told them about the dazzling stairway to heaven he had seen when he was stabbed, with people welcoming him. He knew about death.

"There's a reason you're here," Katie said. "They didn't want you up there yet. There'll be a day when I'm standing up and you'll be walking toward me."

Matt kept wanting to talk about the night of his injury; as he got more upset and angry, both girls tried to redirect his thoughts. Katie wiped away the tears rolling down his face.

"I can't even wipe my tears," Matt said. "That asshole took my whole life from me. I'll never have children."

"Maybe someday," Stephanie said.

After all his friends had left, Ellen and Pat did Matt's evening care because his nurse had called in sick. Then they used the hammock and winch in the Hoyer lift to transfer Matt from his chair to his bed, a vast distance for a tetraplegic person.

Perhaps the lift ran over a wire, or Matt's new air conditioner malfunctioned, or a circuit breaker was overloaded. While Matt was still suspended in the air on the lift, a fire started. As the flames shot up, Pat placed Matt in the bed and rolled it to the opposite side of the room. He tried using a jacket to smother the fire. But the flames began creeping up the wall, so Ellen called the fire department. Matt had to be transferred to Quincy Medical Center—he couldn't remain in the burned basement.

"I've failed him." Ellen wept at losing her son yet again to a medical institution.

Pat feared that he would lose Ellen. She was down to nothing, barely functioning after too many twenty-four-hour shifts. That month, two nurses and a nursing assistant had left because caring for Matt was too difficult.

"Honey, we can't do this." Pat held Ellen as she cried for all of them. "We wanted everything to work for Matt. We put an addition on the house. We really wanted it to work. But it's too much."

6

Riding with an Electric Leg

Patricia Wines

On a sunny day in October 2000, Patricia Wines came home from her job as a dental hygienist, saddled up her quarter horse—named Smashed Rusty for his color but nicknamed Smash—and went for a ride near her home in northeast Texas. A skilled equestrian who stood five-eleven and weighed 150 pounds, Pat had grown up on a ranch with a herd of cows and a stable of horses. Typically she would go to a nearby gym during lunch, but that day she had gone to a restaurant with her husband, Hank. She had been looking forward to some exercise, to galloping and cantering around her forty-four-acre ranch.

Hot and tired at the end of her ride, Pat slowed down near the pond on the way back to the barn. Something—perhaps a rattlesnake—spooked Smash. The horse whinnied and reared up as Pat gripped the reins and hung on with her stirrups. Her experience and strength were not enough to stay in the saddle, though, and the horse threw her off.

In a nearby pen training a colt, Hank saw her in midair and heard her scream. He said that Pat flew fifty feet into the air.

"That's an exaggeration!" Pat laughed as she and Hank told me her story in February 2011, a decade after the riding accident that almost took her life.

"Maybe he was right," I said on hearing about her injuries.

Pat broke four ribs, her clavicle, and her scapula—all on the left side. Scapula fractures are rare because only high-energy forces can break the strong shoulder bone; that fact gives an idea as to the intensity of her fall. Pat's impact was magnified by the hard-packed ground, which was like cement because there had been almost no rain all summer. But the worst injury was the least noticeable: a tear in her left carotid artery disrupted the blood flow to the left half of her brain and caused a stroke.

Hank raced the two hundred feet to her side. "It was the longest distance I've ever run," the tanned, well-built rancher told me. "Pat was groaning, barely conscious. I turned her onto her side because blood was dripping out of her mouth."

He paused at the memory of Pat lying on the ground.

"Then I had to leave her alone and run to the house to call the ambulance." Hank held Pat's hand and said, "That was hard."

"You saved her life," I said. "Pat's lucky to be alive."

"That's what everyone tells me," Pat replied. "One of my doctors said I survived because I was in good physical shape."

"Most people wouldn't have made it," I said.

"They found out about the stroke when I began to wake up and couldn't move my right side," Pat said.

"That was a very bad injury," I said. "The left carotid artery supplies most of the left hemisphere."

Pat nodded. "An angiogram showed almost no blood flow in my left brain."

The nearest town was Jefferson, Texas, which had a population of two thousand and was about thirteen miles away. But Pat had to go to a trauma center with an ICU. Because of the isolated location of their ranch, the ambulance took almost an hour to reach her. Hank held her hand and occasionally stroked her blond hair.

"She was conscious, but the words were just random," Hank said. "The worst part for me was finding out her lung had been punctured by the broken ribs. The doctor had to cut a one-inch hole and stick a tube in her chest."

For the first three days in the intensive care unit, Pat could not move her right arm or leg. Her speech was also affected. When asked to identify visitors, she named them with numbers. Her stroke had been caused by the abnormal blood flow in the left brain, which is where language resides. It resulted in abnormal communication—aphasia—as well as paralysis.

"My husband was 4210, and my daughter was 3640." Pat laughed. "My daughter wished she had written them down, because everyone had a special number." They were random numbers, as I discovered when I asked if any of her name-numbers were related to birthdays or other personal information.

On the fourth day, her older granddaughter came to visit, and a nurse asked her to name the girl. Pat correctly responded, "Chelsa."

After being stabilized, Pat was transferred to Good Shepherd Medical Center in Longview, Texas. At first, the therapist had to sit behind Pat for support because she couldn't remain upright. She could not move her right leg, and her hand was curled up, useless.

Her therapists told Pat's family—which included Hank; her mother; her children, Kamace and Sonny; her son-in-law Stan; and her grandchildren, Chelsa, Madeline, Elijah, and Jackson—that recovery would be a challenge. They feared they might have forever lost the Pat they knew, who was an active, cheerful, and loving person. They would have to accept her disability. Severe strokes result in hemiplegia, but she was triplegic due to her paralyzed right side and her immobilized left arm, which was in a sling to heal her clavicle and scapula. Only her left leg was spared.

Every day, Hank came at six in the morning and left at nine at night, staying to help with her therapies and feed Pat her meals.

"She was like a little bird," he said. "I was so worried that I fed her everything on the tray, so she actually gained some weight in the hospital."

"I worked it off," Pat said with a tolerant smile at her husband.

"Yes," I said. "You look like you're in very good shape."

On arrival at the rehabilitation center, Pat could not use her left arm for weight bearing with the walker, bathing, dressing, or toileting.

"I was very impaired," she told me. She didn't want to go into details of her disabled condition, so I didn't ask further questions. After years of treating patients with severe strokes, I could imagine how Pat might have felt. I have seen patients struggling to eat, open packets of sugar, wash their faces, or put on makeup. I have seen them get tangled up in

their shirts or almost fall over while trying to place socks on their paralyzed feet.

"It must've been terrible," I said.

"Actually, I was lucky," Pat said. "I didn't realize that I was in such bad
shape. I was just happy to be alive."

I tried picturing Pat during her rehabilitation; I've had many
patients with strokes and many with traumatic injuries but rarely the
combination.

"I bet your relatives were shocked when they saw you paralyzed and
disabled," I said.

"Yes, they were," Pat said. "My dad and sister drove down from Utah
while I was in rehab. My dad is six-foot-six, 280 pounds, tough as nails—a
real cowboy. But when he walked into my room and saw me, he cried.
That was the first time I ever saw him cry."

Most of us take movement for granted, but it is a complex process that
involves many brain structures. Some general principles will explain
Pat's struggle with paralysis. First, the nervous system is hierarchic.
Movement starts at the highest level of control in the cerebrum, which
accounts for more than 85 percent of the brain's weight. About the size
of a large grapefruit, the adult brain weighs about three pounds and
contains a hundred billion neurons. Upper motor neurons in the brain's
outer layer, or cortex, cross over to the opposite side; they travel down
to control the lower motor neurons in the spinal cord and cause muscle
movement (see figure 6.1). In both strokes and spinal cord injuries, the
upper motor neuron is damaged, typically along with nerves that control
functions such as sensation and balance. The motor cortex has distinct
areas for controlling each body part; the area for the leg is deep within
the brain, but the area for hand movements is on the surface. A major
artery that nourished the right side of Pat's brain had been damaged, so
her entire left side was weak.

Another principle is that the brain is functionally organized. For
instance, the interconnected clusters of neurons called the basal ganglia
help coordinate movement. Pat's coordination was abnormal, probably
due to injuries in this area of her brain.

A third principle is that brain structures interact and exchange information. This is evident in the connections between the basal ganglia,
the motor cortex, and the thalamus, a relay station for receiving and

sending sensory information. Pat's stroke was so large that it impaired her awareness of her right side, which also affected her ability to move. Most people do not have to look at a leg while walking in order to know its position and to move it effectively, but Pat did.

Millions of neurons are involved in movement. For instance, riding a horse requires planning in the frontal lobe of the brain's outer cortex, which sends a stream of neural signals to initiate a chain of movements. After integrating the information from other motor and sensory areas, the motor cortex sends electrical impulses via the upper motor neurons to the lower motor neurons in the spinal cord, which in turn transmit the signals to the muscles. At its best, the movement is a dance in which a rider becomes part of the horse, trusting the body's wisdom, confident in a lifetime of training that lights up the brain's circuitry.

Before she could go back to riding, Pat had to improve her balance. She could not even sit upright on her own. The therapists asked her to pick up cones and stack them with her right hand. Only after six weeks, when her broken shoulder had healed, was she allowed to attempt stretching and range-of-motion exercises with her left arm. During the eighth and last week of her stay, Pat walked down the hallway with support from three therapists—and covered a distance of about ten feet. There was almost no movement of her right leg, so one therapist braced her knee while two others held her ankle and trunk. At the end, a wheelchair was waiting to bring her back.

"It was very hard work, like nothing I've ever done on the ranch," Pat said. "But I did it. One of my doctors said I'd never walk again."

As she left the rehabilitation center, Pat's therapists told her about their reaction after finishing their initial evaluations. They had walked out, closed the door, looked at one another, and said, "Hopeless."

Pat laughed, thanked them for their honesty, and said she didn't feel hopeless anymore. "After rehab I was in a wheelchair and needed help with transfers, but they sent me home with a walker."

"That was a good sign—they were thinking ahead," I said. "They no longer thought you were hopeless."

"No, they didn't. In fact, they said I'd walk again." Pat said she was glad to return to their home with its wraparound porch, where she often sat and looked out at the pine and oak trees reflected in the nearby three-acre pond. "But when I left the hospital I *was* helpless. I couldn't bathe or dress myself. And my left shoulder was tight and frozen and painful."

"That's a tough phase. You had home services, I assume?"

"We live in a remote area of Texas, so there's no home care. Hank and my mother took care of me. She stayed with us for three years. That's how long it took for me to get better."

After her inpatient rehabilitation, Pat went to see a neurologist and told him, "My foot just hangs down." Her toes were curled down, and she couldn't lift up her foot. This problem is not uncommon after a stroke, due to weakness of the leg muscles, shortening of the heel cord, and abnormal tightness, or "tone." Also known as a "dropped foot," this condition limits toe clearance while walking. The body tries to compensate by hiking up the hip in order to prevent the foot from dragging, which increases the risk of falls. Some patients also swing the leg out to the side in a motion known as circumduction. Both techniques require much more energy than normal walking.

For foot paralysis, an obvious and ancient treatment is to use a brace that compensates for the brain's inability to send a command to lift up the foot. Pat was able to bend and straighten her knee fairly well; most of the tightness and weakness were at the ankle and foot. So she got a plastic leg brace three months after the accident.

"It made my ankle more stable," she said. "But I walked abnormally. I had to jerk up my right hip to swing the leg forward."

Anyone who has worn a cast or a brace for a foot or ankle injury knows that the loss of ankle motion is disabling. It is especially difficult to go uphill and to climb stairs.

Because of her severe stroke and damaged nerves, Pat had abnormal sensations in her leg and foot. This was another problem with the brace, which caused a burning and tingling sensation. Her brain now perceived sensory signals in an intense and painful manner.

But Pat was determined to walk again. She started an outpatient rehabilitation program in Marshall, a town twenty-four miles away. First, the physical therapist placed her on a balance board, which has a curved base. Pat swayed back and forth on it while the physical therapist stood by to prevent a fall.

Her therapy appointments were initially three times each week, then twice each week for three weeks, and finally once each week for two weeks. Pat also had to be trained to place her foot inside the plastic brace, which fit inside her shoe. (From helping patients with this procedure, I know that it is not simple.) With the device, Pat started walking again. At first, using a walker with four legs was necessary. Her gait pattern was awkward, because her ankle was fixed at ninety degrees and forced her knee to bend.

In addition to retraining Pat with her mobility, her therapist started a strengthening program.

"By the end, ten weeks later, they had me on weight machines for the arms and legs," Pat told me. "At the beginning, I needed help just to get into the machine."

For some arm and hand exercises, she started out with just one-pound weights. Pat did bicep curls, pulley exercises, pull downs, and leg presses.

"Did you do any aerobics?" I asked.

"Yeah, they put me on a treadmill, no incline and the slowest speed, about one mile per hour. We had a machine at home, and my husband got me some weights. So now I'm a dedicated weightlifter." She laughed. "Every day I get on the treadmill, and I lift weights."

"That's better than me," I said. "I only use my cross-trainer four times a week. Have you been able to maintain your weight?" Having seen patients steadily become heavier after becoming disabled, I knew this was a difficult battle.

"After the stroke I gained seven pounds."

"That's it?" I said. "You're very self-disciplined."

"If I take a break, I lose strength and gain weight," Pat said. "So I have to exercise to maintain myself. It would be so easy to quit." Pat shook her head and smiled. "And it seems easier every year. Next year I'm going to be sixty."

Her black toy poodle, Cowboy, started barking in the background.

"A thunderstorm's coming," Pat told me.

"A rough and tough little dog," I said. "Does he live up to his name?"

"Yes, we call him that because he chases the cows," Pat said.

"So you're real ranchers."

"We are, but a year ago we sold all but two of our herd."

"You still take care of the cows?" I asked Pat.

"Sure," she replied. "It's part of my therapy."

Pat used the plastic brace for almost five years after her stroke, but it had to be replaced twice. It got too big as her leg lost muscle mass due to weakness and limited walking. Pain was also a factor.

"My foot's very sensitive, and the brace was irritating. I could only wear it for thirty minutes. When it was off, I was just sitting around."

Her gait and balance remained abnormal; even with the device, Pat had to hike up her right hip to prevent her foot from dragging on the ground.

"The right side of my body was mostly deadweight," she said. "Walking was a chore. I couldn't walk and swing my arms at the same time. It felt so unnatural."

Worst of all, the awkward and laborious movements made her fall twice each week. After she hit her head on a doorframe, which injured her left eye and impaired her vision, Hank rarely left Pat alone. He worried about her suffering another serious injury. Now, if Pat looks at a clock, she cannot see the hands in the center because part of her visual field has vanished.

At the end of 2005, Pat saw an ad for the NESS L300 Foot Drop System in *StrokeSmart* magazine, a publication of the National Stroke Association. When she called the maker of the system, Bioness, in California, Pat learned that the device was not available in Texas. She called regularly for eight months and was thrilled to hear the L300 was finally coming to her state. But the nearest location where she could obtain and learn to use the device that would replace her plastic brace was 230 miles away at Texas Health Presbyterian Hospital in Plano, Texas.

"That's quite a ride," I said.

"I was determined," Pat laughed.

For ten weeks, Hank drove her to Plano, where they stayed overnight so that she could have an afternoon therapy session and another one the next morning before driving home.

"The motel owner heard our story and gave us a discount," Hank recalled. "Good guy."

Pat learned that the device has three components that communicate wirelessly (see figure 6.2). The gait sensor is a transmitter attached to a pressure-sensitive electrode. When the heel lifts off the ground, the transmitter sends a signal to the cuff around the upper leg. The cuff has electrodes that stimulate the peroneal nerve to lift up the foot by contracting the front leg muscles. A small microprocessor in the cuff coordinates the electrical impulses with walking to bring up the foot when the heel leaves the ground and the leg swings forward. This sensitive miniature computer adjusts the nerve stimulation to help the patient with uneven surfaces and stairs. The leg cuff is designed for patients with strokes, as they can put it on and take it off with just one hand. A handheld control unit allows the patient to turn the system on and off and adjust the stimulation in case of fatigue or uneven terrain.

In contrast to the white plastic brace, the L300 is barely noticeable as a cuff around the upper leg. Most important, the system improves the patient's gait pattern, reduces the risk of falls, and allows for better mobility outside the home. There is also a clinician's programmer that lets the physical therapist try out different settings for the nerve stimulation and ensure the optimal pattern of walking. These adjustments can be done even while the patient is using the device.

More than four decades earlier, a specialist in rehabilitation medicine named Wladimir Liberson had published his research on a similar device. Electrodes on the skin stimulated the peroneal nerve at the knee, which was triggered by a heel switch inside the shoe. When the heel left the ground to swing the leg forward, the nerve was stimulated and sent impulses to the muscles that lift up the foot. Once the heel landed on the ground, the heel switch sensed that there was no need to lift the foot, and the nerve stimulation was turned off. Bioness used the same principles but incorporated wireless controls. Tel Aviv University researchers Jeffrey Hausdorff and Haim Ring studied the L300 in patients who still had foot drop years after their strokes. Asymmetric walking occurs after a stroke because the weak leg spends less time bearing weight and more time in the swing phase. Due to this uneven weight distribution, as well an inconsistent walking rhythm, the risk of falling is high. The researchers intended to find out whether the device would improve balance, the gait pattern, and the risk of falls.

After patients used the L300 for eight weeks, their walking was faster and more symmetrical. Of greatest importance to Pat was the fact that the frequency of falls decreased markedly. Most patients were able to operate the device after being trained and to correctly place the system's components with one hand. They stated that it was safe, comfortable for extended use, and not a problem in social situations. All except one wished to continue wearing the L300. All of them increased their physical activity and had more confidence with inclines and uneven surfaces. As to problems, some stated that the flashing lights during stimulation had elicited comments from strangers and that the electrodes needed replacement every two weeks. All in all, the Bioness electric brace benefited patients after strokes.[1]

Pat realized that the L300 could make a real difference in her life as a stroke survivor. She started off using the device for short periods in order to ensure optimal placement of the electrodes in the leg cuff.

"When I first put on the L300," Pat said, "it was like when you bump your elbow and get a zap, but not as intense. You can raise or lower the intensity. Now, after all these years, I don't even feel it."

Once again, she had to relearn how to walk, but this time in a more natural pattern. Her therapist taught the proper ways to shift her weight and make the most of her remaining strength and balance.

"I'd gotten into some bad habits to clear my foot with the plastic brace," she told me. After learning to use the L300, she said, "I walk more normally. I'm free from walkers, canes, and my brace. And my fatigue is gone."

Her endurance also improved. Pat gradually increased her treadmill to the highest incline and a speed of 3.2 miles per hour, which is a fast walk. At present, she does this for thirty to forty-five minutes, five times each week.

Since using the L300, Pat has had only two falls; the first one was when she tripped over Cowboy, and the second was when she was barefoot, without the L300.

"A couple of times I almost fell but caught myself. The L300 picks up my foot, and I can recover."

With more normal mobility, Pat now walks on a variety of surfaces, hiking around the vast expanse of her ranch and strolling along the seashore during family vacations at Pensacola Beach, Florida. With the L300, she no longer needs a ramp for the eight steps up to her porch. Her tendency to fall with the plastic brace had kept her away from social settings, especially ones with crowds, but now she has enough confidence to go shopping at the mall and attend soccer games and church events. Pat had essentially given up the idea of dancing again, but even that is possible.

In a follow-up study of the L300, the researchers at Tel Aviv University found improvements in physical and social adjustment to strokes.[2] Patients improved their walking speed at two months and even more at one year, probably because their gait patterns were more normal and efficient. Pat's experience with the L300 occurred before this research was conducted, but it is likely that her physical and social improvements were noted by Bioness and contributed to the design of the research.

Why did Pat achieve such dramatic success—five years after her stroke—with the NESS L300 Foot Drop System? In their long-term study, the Tel Aviv researchers suggested a few possible mechanisms, both in the affected limb and the brain. First, electrical stimulation of the peroneal nerve, which lifts up the foot, improves the strength of the leg muscles and decreases the tightness or spasticity that often occurs after a stroke. I have seen my patients with strokes achieve a new level of mobility once these improvements occur. Walking greater distances leads to better cardiac function, endurance, balance, and lower-limb strength.

Recent research has shown that even years after a stroke, the damaged brain receives information about the patient's recovery. For instance, electrical stimulation of the pharynx leads to increased activity of the brain's motor cortex; this research by Shaheen Hamdy and his colleagues is the basis for functional electrical stimulation (FES) in the rehabilitation of swallowing deficits after a stroke.[3] Alain Kaelin-Lang and his fellow researchers proved that electrical stimulation of the ulnar nerve at the wrist increased the electrical excitability of upper motor neurons in the brain's cortex.[4] And electrical signals at a finger muscle controlled by the ulnar nerve also increased. This study showed that nerve stimulation increased the electrical activity of the brain's motor neurons *and* the muscles they activate.

Another explanation for Pat's success is that repetitive movements, such as lifting up the foot when walking, are necessary to regain lost functions after a stroke. This is a fundamental principle of neurologic rehabilitation, but the repetitions have to be natural and efficient, as close as possible to the healthy state. A plastic brace is helpful but not natural because the immobilized ankle results in abnormal forces at the knee. Through thousands of repetitions with the more natural walking pattern of the electrical brace, the brain relearns how to direct the body to walk in a fluid manner, efficiently and safely.

To study the impact of intensive arm and hand therapy, Joachim Liepert and his colleagues used magnetic stimulation of the scalp in patients whose strokes had occurred at least a year prior.[5] First, the researchers determined the area of the brain that controlled the hand. Intensive hand therapy increased this area and led to better movement and function. In a similar manner, the Tel Aviv University researchers suggested that the L300 provided repetitive practice through a variety of mechanisms. Because the device does not immobilize the ankle, in contrast to the plastic brace, it requires patients to work harder to stabilize their ankles and knees. Their movement becomes more fluid and less awkward, reducing energy consumption and increasing stability. As a result, they are more likely to walk greater distances, which leads to more repetitions and theoretically more effective relearning and rewiring of the brain.

Pat's greatest ambition after the stroke was to walk out to her horse in the pasture and ride it. After five months of using the L300, or almost a year after her stroke, Pat achieved her goal. She walks on rough nature trails without falling and can step over a stream without getting a mud bath. She is careful to protect the device from water, for getting it wet would nullify the warranty and insurance. But in rare emergencies, such as having to feed her horses in the rain, Pat zips up a plastic bag around the gait switch near her foot.

"Some people ask why I let her ride again," Hank said. "I tell 'em she loves horses, she was raised with horses, and I'm not going to take that away from her. People don't stop driving a car after an accident. It's the same for a horse."

"Hank helps me get on and off the horse," Pat said.

Then we discussed another reason for Pat's success with the Bioness device: it helped her sensory awareness of her right leg. Her left-sided stroke had damaged the sensory cortex on the outer surface of her brain, so she wasn't always aware of what her right leg was doing. On top of her weakness, this sensory deficit worsened her balance. The electrical stimulation of the L300 provided sensory input that certainly made her more aware of her leg and probably helped renormalize her perception of different sensations.

"I'm more aware of my right side," she told me. "It feels like part of me again."

Unlike Anna Iacono, who functioned quite well with a plastic brace and a cane, Pat did not benefit from a conventional brace. In fact, it compromised her balance and normal pattern of walking, which led to falls and a serious eye injury. The plastic brace made her more tired because using it required more energy. It also irritated her hypersensitive foot and increased her level of stroke-related pain.

Pat's fractures, stroke, job loss, and disability led to severe depression, and in 2001 she received treatment with the antidepressant Effexor. Until her stroke, she had been a happy, positive person and had never used any medications for depression. After starting the L300 in 2006, she was able to stop the medication by the end of the year. Pat also told me another way in which the L300 changed her life.

"After my mobility got better and I wasn't afraid of falling, Hank talked me into traveling more. So we bought a motor home. It's a beige and brown Dutch Star, the size of a large bus. I can handle the two steps to get inside and two steps to the main level."

Pat uses her electric leg while camping, but she always brings along her old plastic brace and cane as a backup. Recently she was on a boat and misplaced her L300, which she stores in a waterproof bag. With her cane and brace, she trudged up a hill, hobbling along until she reached her motor home. A friend saw this feat and said, "You don't set down and quit. You regroup, smile your big smile, and keep on trucking." Pat responded that she had learned to adapt.

"With my patients," I said, "I find that the ones with the most positive attitude often do the best."

"No matter what life throws at you, there's always a ray of sunshine," Pat said. "By the way, the next day, we found the bag stuffed in with the water-ski equipment in the hull of the boat!"

"You don't water ski, do you?" I wouldn't put it past Pat, but it's not a sport I'd recommend for most patients who have had strokes.

"I don't ski," Pat laughed. "I'm the cheerleader."

Pat was lucky in that her insurance covered 80 percent of the cost of the device. Hank is retired from the military and has TRICARE insurance. To help get the device, Pat's doctors and therapists wrote letters extolling the L300's benefits. Hank and Pat were surprised because the reimbursement came more than a year after they bought the device.

"It was worth it because it made me happier. I feel much better when I'm active," Pat explained.

"Your exercise schedule is pretty remarkable. Most people who have strokes don't do this," I said.

"I was fairly active even before the stroke," Pat said. "My husband doesn't exercise at all. But he cheers me on. He gets on the computer and surfs the net while I exercise."

"Well, you'll convince him someday."

"I've gotten him to walk with me, so that's a start."

Pat and Hank enjoyed walking along the lake, which was only a hundred yards from their beige house with red shutters. Occasionally they brought their fishing rods to cast out for bass and catfish, which could grow up to ten pounds. The water was tempting in the summer heat, but the lake wasn't safe for wading or swimming.

"The banks are muddy and slippery," Hank said. "And we don't want to swim with the water moccasins."

We finished our discussion by talking about a photograph of Pat with her granddaughter, Madeline, and her brown mare, Comet (see figure 6.3).

"Madeline tells me to put my knee on so we can go out and play," Pat said.

In the photo, Pat holds two red feed buckets that are filled with grain, and she seems to have a good grip on them. Her right hand had recovered most of its strength, but her fine-motor function and sensation were still abnormal.

"If I look away from the bucket too long, I'll drop it," Pat said.

Next to her in the image, Madeline smiles while stroking Comet's white-streaked forehead. Pat grins, her blond hair blown by the wind, the L300 invisible under her jeans.

7

Mt. Sinai's Medical Commandments
and the Wheelchair Superman

Matt, Late 2002 to 2003

Because of the nursing shortage, Matt could not stay at home and was transferred in July 2002 to New England Sinai Hospital and Rehabilitation Center for long-term care. Hidden away in the woods on a few acres in Stoughton, fifteen miles from the Nagles, the large two-story building had a pebbled gray-brown stone facade. Near the marble-floored entrance were wheelchairs for visitors, a potted fig tree with glossy green leaves, and smaller pots of pale pink orchids. A lobby wall was decorated with a metallic sculpture of a brightly shining sun, its rays piercing a bronze tree of life that grew out of the wall. Stained-glass windows looked out onto a sunbathed garden of rhododendron and juniper bushes. A hallway displayed a wall relief sculpture in Hebrew lettering: HEAR ME, O GOD, AND I SHALL BE HEALED IN THIS PLACE OF HEALING OF BODY AND SPIRIT.

An athletic overachiever like his son, Pat hated to fail Matt, but New England Sinai was the best place for his complex medical needs.

"I don't want to be in a hospital," Matt said. "I want my own apartment."

"You know, buddy, we'll get you an apartment. And in two days you'll call your mom to come and take care of you. But she has to work; she can't do that."

Matt just nodded at his father, who couldn't tell if this was a sign of resignation or frustration or anger, or a blend of those emotions.

Later, Ellen tried to explain the situation to Pat.

"He thought if he got his own apartment, his friends would be over and things would be more normal."

"But most of his friends have . . . gone," Pat said. "From high school, there's only Katie and Danny."

Danny Taylor hadn't been one of Matt's best friends in high school; he wasn't a heavy partier or a star athlete. Yet after Matt's friends stopped coming—for reasons ranging from lack of time to discomfort with his paralysis—Danny continued visiting him. He accepted the mission of helping a severely disabled friend who could give back nothing except his thanks.

"This kid's unbelievable," Ellen said.

"He has a great heart," Pat said. "Danny's pretty special."

"Do you know which friend never visited Matt?"

Pat frowned and shook his head.

"The kid in the beach fight," Ellen said, "the one Matt went to help."

The Nagles agreed that, though New England Sinai would take him out of his comfort zone and potentially away from his friends, it would be the best place for Matt to receive care for his medically complex condition.

At the long-term hospital, Matt's physician, Dr. Brian Bloom, came to realize that his patient hadn't grown up before his injury—or afterward. Matt had not yet accepted the life of a patient, with its many limitations and frustrations. For instance, just going along with the breakfast schedule was a problem for Matt, who enjoyed late-night television and often slept in.

And there were other struggles. Matt wanted narcotic medications more often than was medically ideal. These requests were probably related to his severe shoulder pain, which felt as though the entire trauma to his body was focused there, like sunlight burning through a magnifying glass. But the staff did not want to overmedicate him and cause adverse side effects, such as confusion or a bowel impaction.

So Matt and Dr. Bloom would get involved in complex verbal dances. Matt wanted some medications at certain times or not at all. Or he wanted to have a diet that included pizza, potato chips, and soda, in spite of his doctor's concerns about weight gain and the risk of diabetes. Dr. Bloom realized that there were only a few things in his life that Matt could control: when and what he ate, when he had his bowel and bladder care, and when he had medical appointments. It was a balancing act to provide medical care and to allow Matt the control he needed.

On rare occasions, Matt desired the ultimate control over his life: his death. When Danny visited his friend early on at Sinai, Matt asked him to shut the door and spoke quietly, his sandpaper voice forced through the tube.

"Danny, unplug my vent. I can't live like this. Come on, do it for a friend."

"Matt, I just can't." Danny shook his head and turned away.

"Unplug it and walk out. They won't notice—until it's over," Matt whispered.

"Sorry, Matt, I know you're going through hell, but I can't."

"Fine! I'll get a real friend to do it."

Matt refused to speak to Danny, so he left. He told the Nagles that Matt didn't feel life was worth living.

During a visit with other friends, Matt again talked about disconnecting the ventilator.

"I've already lost my life," Matt said. Pat heard about it from someone who thought a friend might oblige Matt.

"If anyone pulls the plug, I'll go after that person for murder." Pat pointed his finger at the informant friend. "You tell everyone I said that."

When Matt heard about this threat, he told Pat, "Mind your business. You've always got to play cop."

But Matt didn't follow through on his suicidal thoughts. His parents had sacrificed so much for him, so he held on for Ellen and Pat.

The Sinai staff members learned to make sense of Matt's behaviors and to tolerate them. One of Matt's favorite nursing aides, Kayla, did her best to understand and comply with his idiosyncrasies. At first, Matt was intimidating and demanding. She had never had a patient ask for one face cloth for each body part. Paradoxically, he was fussy about certain activities and casual about others. For instance, Matt wasn't good about dental hygiene, and he sometimes put off brushing his teeth. But he was always careful about what he wore. His favorite combination was a pair of sweatpants and a Patriots jersey, finished off with athletic socks,

Adidas sandals, and a cap chosen from his collection of Red Sox, Patriots, Celtics, and Bruins gear.

In spite of his condition, Matt loved life, Kayla thought, and retained his sense of humor. He often joked and made her laugh even as she catheterized his bladder or struggled with his clothing. In the wheelchair, Matt liked to be covered with a Patriots blanket for comfort; he couldn't feel the cold because of his sensory loss. If a female friend was coming over and he wanted to look good, he would ask Kayla to put gel in his hair. But if he felt down, he wore his blue hospital top and pants.

Matt was usually up at ten in the morning, and Kayla fed him his favorite breakfast: oatmeal, peanut butter, and toast. If he had a craving for Rice Krispies, he polished off three boxes in one sitting.

After Kayla washed and dressed him, he was placed in the wheelchair with the Hoyer lift. Getting Matt into his chair was complicated, requiring someone for the lift as well as a respiratory technician in case his vent malfunctioned. Secured in the hammock of the lift, Matt was winched up in the air, rolled over to the wheelchair, and let down. He had to be positioned properly because his paralysis and lack of sensation greatly increased his risk of a pressure ulcer. Now, Matt could control the motorized wheelchair with the sip-and-puff tube mounted near his mouth. When he exhaled (or puffed) forcefully, the change in air pressure sent the chair forward; forceful inhalation (a sip) sent the chair backward.

Lunch around two o'clock suited his schedule the best, and Kayla made sure that he was her last patient, in order to give him all the attention he wanted. Chicken fingers were his favorite meal, with macaroni and cheese a close second. And Kayla always gave him chocolate chip cookies for dessert.

Matt loved Ellen's homemade beef stew and saved that treat for days when the hospital menu was not to his liking. If he missed lunch or dinner, or if Kayla wasn't around, Matt ordered pizza. To satisfy a craving for fast food, such as a cheeseburger, he ordered out late at night; the staff usually indulged him by dialing the phone number. (Early in his residence at Sinai, Matt didn't get along too well with the staff. But as his friends faded away or left, he developed a more friendly relationship with the nurses.)

Staff members continued to help Matt with phone calls, but then someone overheard him betting on sports. Now ineffective on the field and rink, Matt found that gambling was the closest he could get to the thrill of the action, the joy of the body striving to its limits. The hospital couldn't allow its staff to be gambling assistants, so his calls had to stop.

Matt turned to other entertainment. When Danny visited, they sat by his truck, soaking up the sun, listening to the radio, and chatting. Matt somehow stashed away cigarettes in his room and persuaded his friend to bring the red-pack Marlboros outside so that he could have a smoke. Even though Danny disapproved of the activity, he saw it as an escape from the hospital, a treat that was more necessary since the stabbing. If he didn't help with the cigarettes, Matt would get angry and say, "Look at me! Come on, what else do I have? What else?"

If the weather was bad, they stayed inside and watched sports on TV or movies on Matt's laptop. Danny was often surprised at how quickly Matt obtained new films, usually bootleg recordings from the movie theater. Occasionally a person walked across the screen or they heard moviegoers eating popcorn, but that was the only way Matt could enjoy new movies. The logistics of getting his motorized wheelchair and portable ventilator into a theater were difficult. Besides action thrillers, he liked horror films such as *Friday the 13th* and the *Saw* series.

Social events also distracted Matt from his condition, but travel was complicated. With a great deal of coordination and preparation, Matt attended the Cape Cod wedding of his cousin, Kim, and served as an usher. The Nagles hired his former caregiver, Gina, to watch over him and ensure that his ventilator wouldn't fatally malfunction. They noted that the two were very fond of each other.

"Maybe she was more than just a nurse," Pat mused after the wedding, "but I figured it can't—it physically can't happen. Even with Viagra."

"The way she took care of him!" Ellen said.

Matt could have had a relationship with his caregiver. The Nagles were unaware that some sexual function is still possible after a spinal cord injury. One of his nurses also wondered if Matt was physically capable of a sexual relationship. When she came across Viagra in his room, it had to be taken away. Matt occasionally needed nitroglycerin for dangerously high blood pressures due to autonomic dysreflexia, and taking both medications at the same time could cause his pressure to plummet.

Matt's care at New England Sinai required meticulous attention. His bowel program every other day involved a Fleets enema followed by stimulation with a gloved finger for reflexive emptying. A nurse emptied his bladder with a catheter every four to six hours. A condom catheter and a bladder bag were necessary, because he occasionally voided on his own. Matt had accepted that his life was being measured out in plastic urine bags. An open pressure ulcer on his coccyx was covered with an occlusive patch dressing called DuoDERM, and Matt was regularly turned side to side to prevent it from worsening. With his chest muscles paralyzed, Matt could not cough. A coughalator machine was plugged onto Matt's tracheal tube to force normal saline and air into his lungs; then the secretions were suctioned out. If he needed help, Matt moved his head against a pressure-sensitive pad near his pillow and triggered the call light. He was settling into a routine at New England Sinai. He was learning the patience of paralysis.

Katie Perette realized that Sinai was a haven for Matt, but it was also a prison. She had remained a steadfast friend, comforting him, enduring the sadness of Matt's electric wheelchair, and not letting him push her away as he had so many others. So when Matt moved there, she decorated his room so that it felt more like his own place, a refuge and a lair.

Katie also organized a fund-raiser for Matt and started a letter-writing campaign to enlist the Patriots and one of Matt's favorite bands, Aerosmith. (Matt especially liked the video of the rock-rap version of the band's "Walk This Way," performed with Run-DMC. It featured the two groups engaged in a musical duel in adjacent studios until Aerosmith's Steven Tyler crashed through the partitioning wall for a duet.) Knowing about Matt's interest in stem cell therapy, Katie contacted a network of people committed to that research, and she even tried to bring Hans Keirstead to the event.

To thank Katie, Ellen wrote:

> You are an amazing woman, Katie. You've taken on an enormous undertaking, and Matt is truly blessed to have you for a friend. Best of luck at the fund-raiser meeting next Wednesday. Matt says hi.
>
> Love,
> Mrs. Nagle

Katie hated going to Matt's facility and often cried after leaving Matt in the small room where he could barely move his wheelchair. She felt sad that he was in a wheelchair all day, only able to watch TV shows for entertainment. The money from the fund-raiser would be great for covering expenses, but it might also lift his spirits. She wanted Matt to buy a robot system or a computer or whatever he wanted. Katie's goal was ten thousand dollars, the cost of a voice-activated robot system that could help Matt with basic activities, such as feeding himself. To surprise Matt, she invited him the day before the event, which was being held at the Knights of Columbus hall. At first he didn't want to attend, but Katie talked him into going.

Katie's preparations included special T-shirts, hot pink for the girls and gray for the boys, imprinted with NAGLE and 21, Matt's football number. Her family, restaurateurs in Weymouth, donated the food. As Matt entered in his wheelchair, the girls lined up on one side and the boys on the other, totaling more than a hundred high school friends. Matt started crying.

At the end of the line, Katie was waiting for him.

"I love you, kid," Matt said as she bent down for a kiss.

Katie told Matt that he had to thank them all.

"Everyone's by your side," she said.

She had never seen him so happy. At the end, his friends formed a large circle with Matt at the center, and they sang "That's What Friends Are For." The event raised thirteen thousand dollars. While handing him the money, Katie said that everyone there loved him.

<hr />

Many people in the greater Boston area felt for Matt and the Nagles. When football season started, Tom Brady, Tedy Bruschi, Adam Vinatieri, Troy Brown, and some other New England Patriots signed a jersey with Matt's name and number on it. This gift was arranged by Danny's friend, a personal trainer for the Patriots. The team went all out for the former football star. Bob Kraft gave Matt and Danny luxury box seats and locker-room passes for a game against the Vikings.

Help also came the day after the Boston Marathon, when Pat opened the front door to find an athletic elderly stranger named John O'Brien. He handed him a thick envelope. Inspired by Matt and aided by generous sponsors, he had raised over five thousand dollars by running the 26.2-mile race.

Athletic events were natural fund-raisers for Matt. Three bike-a-thons were organized by Jerry DiLillo, Ron Pollara, and other Weymouth residents, with help from the *Patriot Ledger*, the *Weymouth News*, and the *Boston Globe*. In support of Matt, a hundred people bicycled up through Hingham to Cohasset and then on to Hull, the strip of land from which one could see across the silvery blue water to Boston Harbor. The ride was called "Miracles for Matthew."

On Sundays, Matt went home. As his anticipation built during the middle of the week, he would ask Pat, "What's Ma making? What's Ma making?" One of his favorite meals was shish kebabs, rice pilaf, and salad. He also loved lobsters, so Mike got eight-pound lobsters once or twice each year for a big cookout. Matt was in heaven. Mike parboiled, cleaned, and basted the lobsters with melted butter to enhance their slightly sweet and nutty flavor. When they were grilled, Ellen fed the best pieces to her boy. Matt closed his eyes, savoring not only the aroma of grilled seafood but also the view of their tall pine tree, the new-wood smell of the deck Mike had built for him, and the love of his family.

When some of Matt's friends went to Sinai, they brought movies to watch with him. Immobile, Matt had few options for social activities.

Once Meaghan Murphy asked Matt, "So what'd you do today?"

"What do you think I did? I watched TV."

Meaghan tried not to feel stupid about her question. It is difficult for able-bodied people to understand the universe of those with severe disabilities, and there are many occasions to inadvertently and innocently say the wrong thing. She imagined it was difficult, if not impossible, for him to read the sports section of newspapers or magazines. Who would hold them and turn the pages for him? It was a lot easier to watch the TV.

During another visit, Meaghan saw Matt wearing bright white shoes, so she casually asked, "New shoes?"

Matt gave half a smile and said, "No, Meg. How can I get any of my shoes dirty?"

Although a little disturbed by this response to her offhand remark, Meaghan felt that Matt wasn't being rude or trying to make her feel stupid; he was probably tired of questions from people who had no understanding of his suffocating condition. Meaghan realized that if she were confined to a small, cluttered room all day, tethered to a ventilator by plastic tubes, and forced to drag around her paralyzed body on the wheels of a motorized wheelchair, she probably would have little tolerance for comments that showed a deep ignorance of disability. Although these awkward moments didn't damage Meaghan's friendship with Matt, they didn't encourage visits either. But she tried her best to support him and see him as much as possible. She tried.

Matt's comments to other people, however justifiable, took a toll. Over the next few months, there were fewer visitors, fewer gifts, fewer cards, fewer e-mails, and fewer reporters. Gradually, Matt's universe was getting smaller.

Some people stuck by him. Instead of exchanging Christmas gifts, the neighboring DiLillo family sent a card with a large check for Matt. But as much as his well-wishers desired, no one could give him the gift of health.

Over Christmas, Matt was sent to an acute-care hospital to treat a pneumonia that flooded his bloodstream with bacteria. The Nagles discussed whether Matt should go back to New England Sinai.

"When he first moved there, Matt didn't come out of his room a lot," Pat told Ellen.

"He didn't want to see people in wheelchairs," Ellen said.

"I'd ask him about the person in the next room, and he didn't even know his name," Pat said. "But now he knows everyone on his wing. He knows their sad stories, and he feels for them." The Nagles agreed that Sinai was still the best place for Matt's complex condition.

Paralyzed, his muscles no longer stabilizing the shoulder joints, Matt developed chronic pain in spite of Oxycontin, Percocet, baclofen, Neurontin, Buspar, Celexa, Vioxx, cortisone injections, heat treatments, and range-of-motion exercises. Although he could sense pain at the shoulders, he could not move his arms to a less painful position.

Due to Matt's dependence on a ventilator and the plastic tube in his throat, the scar tissue in his trachea kept coming back, stubborn as dandelion spores. At Massachusetts General Hospital, he endured more laser surgery and the placement of a stent, a hollow tube to prevent scarring. Each surgery caused intense throat pain and left him unable to eat for some time.

After seven surgeries, Matt spoke in a raspy voice: "I never got a chance to thank the doctor. He died a month after the surgery. I hope he can hear me now."

In the institutional setting of New England Sinai, Matt learned how to seal off his anger, tightening the bolts with as much mental force as he could generate. When he became frustrated with his helplessness, when he lashed out at a nurse for not getting a pain pill fast enough or for bringing the wrong breakfast, he learned that even nurses had limits to their patience. But his mother was inexhaustible. As the gusts of his anger blew against her, Ellen was steadfast as a tree that would bend but always return to its upright position.

After overhearing one of these outbursts, for Ellen rarely complained, Pat decided to have a talk with his son. On the way up to Sinai, he stopped off at Dunkin' Donuts for Matt's favorite coffee, a large vanilla with milk and artificial sweetener. The best visits were when he walked in, unannounced, with a hot coffee. Pat knew he wasn't supposed to get him a doughnut, but this was an exception. Matt rewarded him with a smile, and Pat leaned over to kiss his son.

They talked about spring training for the Red Sox while Pat held the cup and Matt sipped the coffee through a straw.

"Buddy," Pat began, changing the subject, "no one loves you more than your mother. She's the easiest person you can hurt in this world. She'll cry and be sad, but she'll come back to you. Be good to her."

The straw fell out of Matt's mouth.

"She loves you more than anything," Pat said. "When you got injured, she was your lifeline—and mine."

Matt then confessed that he had railed at Ellen for allowing him to live, for not yanking the ventilator tube out of his throat. He cried, and as Pat wiped away the tears, Matt promised to treat his mother better. He also agreed to spend some time with the psychiatrist at Sinai, to understand his anger and control it.

His parents also hoped that some kind of employment would help Matt deal with his condition. Ellen sent an e-mail to Cynthia Sweetser at the Massachusetts Rehabilitation Commission:

Last year you spoke with Matt regarding vocational rehab. Matt is very depressed and despondent. It would be tremendously beneficial if you met with him regarding adaptive technology, possible employment/retraining, or school options. He is very willing to speak with you. Matt is about nineteen months postinjury, and I believe he has reached a turning point with regard to "getting on with life" as best he can.

Cynthia responded that she looked forward to meeting Matt. At the end of February 2003, Ellen sent another e-mail:

Is there anything I can do to help? As I said before, he's pretty discouraged. I pray he's ready for your visit, to get started with vocational rehabilitation. Thanks so much for your help.

For reasons he did not wish to discuss, Matt decided he was not interested in a vocational-rehabilitation program and refused to meet with Cynthia.

With almost no control of his own body, Matt could defend himself or attack only through words. Sometimes he shocked his parents and the staff. Ellen realized they had to set limits, to let him know he couldn't talk like that. But he still had limited physical resources, as an aide discovered when she offended Matt and he got angry and spat at her.

"That's an absolute no-no," Pat said. "They won't put up with spitting."

"Don't tell me what to do!" Matt said in a hoarse voice.

"Matt, we pulled a lot of strings to get you into Sinai," Ellen said. "It's the best facility around. If you keep this up, they'll transfer you. They'll put you out on the street."

Matt was silenced by this comment. He turned away from his parents and stared out the window at the tall white pine weighed down by the snow.

"There's nothing we would be able to do about it," Pat said. Unlike Ellen, Pat was less forgiving of Matt's occasional outbursts. "When you want us to come back, you can call."

Angry at both his parents, Matt felt trapped in a facility where most of the residents were elderly—and inside a body that no longer belonged to him or gave him any freedom or pleasure, a body that weighed him down, pulling him deeper and deeper into the ground. During almost a month of despair, the Nagles didn't visit because Matt didn't want to see them. Matt remained alone at New England Sinai, his only companion the rage that prowled inside his small room.

Pat saw that Ellen was distraught and tried to soothe and comfort her.

"We can't let him treat us like this," Pat said. "He's got to call us."

"But he's there all by himself." Ellen held on to Pat as she wept.

"Even if he doesn't apologize," Pat said, "if he just calls for a sandwich, that's enough. We'll go back. But he's got to learn. He's not talking to some kids on the street. We're his mother and father, and we're not going to take it."

"OK," Ellen said quietly. "Maybe it'll help him deal with people in the long run."

As it turned out, that was exactly what happened. In his disabled state, Matt needed to learn how to manage his family, friends, and caregivers.

Matt never had any interpersonal problems with his devoted uncle Rick, who often brought him a pizza or a meatball grinder or buffalo wings. Even though Matt shared his treats with the staff, a lot of the food ended up in the refrigerator as leftovers, for his dining schedule was unpredictable.

The night staff tried to help Matt understand and live with his spinal cord injury. They humored his finicky requests for late dinners and allowed him to stay up past midnight, watching the sports news on ESPN or action movies. Almost two years after he was stabbed, after his anger had driven away many caregivers and some friends, after some tough times with the staff on C wing, Matt channeled his anger into more productive interactions with people. His nurses commented to one another that he was nicer to the staff than he had ever been.

⁣⁣⁣⁣⁣⁣⁣⁣⁣⁣⁣⁣⁣⁣⁣⁣⁣⁣⁣⁣⁣⁣⁣⁣⁣

Every July 3, the Nagles made sure to have Matt with them at home.

"The last thing I want," Pat said to Ellen, "is him lying in bed that day with no one around. My parents were married on the Fourth, and I used to love the day—the whole nine yards. The fireworks, the parks, the grills. . . . Now I'd just as soon sleep through it and wake up on the fifth."

At a cookout, the Nagles prepared Matt's favorite meal: chicken-and-vegetable shish kebabs. Only four people showed up. As for the rest of his friends, their excuses were barely credible. Matt joked and laughed less than at the previous year's event.

Katie Perette commented afterward to the Nagles, "They were all his boys, but they all faded away—all of them. All except Danny Taylor. He wasn't even good friends with Matt before the stabbing, but he turned out to be a real friend."

Ellen sighed, doused the grill, and watched the smoke vanish into the pines.

Katie never told the Nagles that Matt later cried to her, "Where are my friends? Why don't they come? Because I can't have a beer with them anymore?"

"Matt, I can't explain it," Katie said. "They say if you have five good friends, you're very lucky. Most have only one. I've lost a lot of friends along the way."

Matt asked Katie to call a friend on her cell phone, and he said, "Why don't you visit me? I miss you."

"I'll be there tomorrow, with a steak, and I'll bring a couple of the boys."

Late the next evening, Matt called Katie.

"I didn't eat. I waited for the steak. . . . They never came. Dinner's been served. Katie, can you bring up a steak or something else?"

When Katie arrived, Matt asked her to lean over so he could kiss her cheek.

As his friendships withered away, Matt tried reaching out for support through the Internet. In October 2003, Ellen wrote a letter for him to Christopher Reeve, the actor famous for his role as Superman, who was fully paralyzed after a horse-riding accident:

> I would like to start this letter by thanking God for giving me the opportunity to write to you. I would also like to thank my mother, who is typing this letter for me. She's been my biggest supporter and has never left my side.
>
> I would like to thank you, Mr. Reeve, for giving me hope and strength in an otherwise hopeless situation. Your courage,

I'm sure, has inspired many people in wheelchairs with physical disabilities.

Matt described himself as a young man from a Catholic family who was a *Boston Herald* All-Scholastic athlete. Then he related what happened after the attack:

When my parents arrived the doctors told them to prepare for the worst. Three of the lobes in my lungs had collapsed, and I was in a coma and on a respirator. . . . The support I received was amazing. . . . At that point it wasn't clear if I could eat or drink. Then it started to kick in—I was in trouble. I stayed at Shepherd for two months. I was sick from UTIs and blood clots and various infections for over five weeks. I was discharged on October 1. I thank God for what the people at Shepherd did for me.

Matt recounted his problems at the nursing home, where he developed a pressure ulcer:

My mother pulled me out of there and took care of me for five days straight. She's my angel. Finally, we started to get nurses and PCAs, and by the New Year, we were able to fill 24-hour shifts.

In February, I went to Rhode Island Hospital for a phrenic nerve pacemaker so that I could get off the vent. Unfortunately, the stab wound to my neck had damaged my phrenic nerve. It was a big letdown. At that point I was unable to speak, even after three surgeries to my vocal cords in Atlanta. The problem was that scar tissue continued to grow over my vocal cords.

Matt described the one-year anniversary of his stabbing, the huge cookout, the fire, and its aftermath:

My parents came to tell me that the money from my spinal cord fund had dwindled to almost nothing, some of the nurses had quit, and I would be forced to stay at New England Sinai Hospital in Stoughton, Massachusetts. At the time I was very bitter.

When I think back, I realize it was too tough for all of us. It was an unhealthy situation.

One of the things that I learned at Shepherd was to be an advocate for myself. It took a lot of compromising and me speaking my mind, but after about six months, I had established a routine and made friends with the staff. From November to May, I endured three more surgeries on my voice but I still couldn't talk normally. I was battling a deep depression. One day I wanted to die; the next day I could cope. There were so many emotions running through my head that I do not wish upon anyone.

In July 2003, a doctor from Mass Eye and Ear took out the stent he had placed in my throat to prevent scar tissue from regrowing over my vocal cords. That was my seventh and final surgery. I started working with a respiratory therapist by the name of Bob Chase. He was a genius! He was able to trouble-shoot the ventilator so that I could speak.

I still remember the day when he mentioned that he had worked for you. Earlier in the week, some people asked if I had seen you on the *Larry King Show*. They said you could take a few steps using parallel bars. Bob mentioned to me that you were going out of the country to get stem cell therapy. For the first time since the day I was stabbed, my body filled with hope. Bob said that on October 15, you were planning to make a speech to a group of respiratory therapists in Sturbridge. I was ecstatic! I couldn't believe that I might have an opportunity to meet you. I have always told people that if I am ever given an opportunity to walk, I would be willing to go out of the country to get stem cells. Bob mentioned that you went to England, Spain, and Israel for treatments. I can't tell you how much strength, hope and courage you've given me.

Through the wonderful support of the people on the South Shore, friends and family, we have raised a lot of money for my fund. We've established a golf tournament where the proceeds go to the Matthew Nagle Spinal Cord Fund. The money is used to help pay expenses for other people who have suffered a spinal cord injury.

I am now filled with hope and a mind-set that I will walk again. They say if you put your mind to something, you will accomplish it. You are a perfect example of this, and I plan to follow in your footsteps. Mr. Reeve, I am going to ask you the most important question that I have ever asked in my life. If you would be so kind as to please, please, please (in complete confidentiality) tell me the names of the hospitals and doctors that you have worked with outside the United States, it would mean the world to me. Money is not an issue. If you could share that information with me, I would be indebted to you for the rest of my life.

I want to tell you how sorry I am that this tragedy happened to you. I do not wish this on my worst enemy. I think it's fair to say we've both been to hell and back. In closing, I want to say, God bless you, your family, and everyone out there with any type of handicap. You are my hero, Mr. Reeve. You are a bigger man than Superman could ever hope to be. Thank you for reading my letter. I look forward to hearing from you soon.

Love,
Matthew Nagle

In spite of heavy doses of many medications, pain kept gnawing away at Matt's shoulders. Worst of all, it was unpredictable, and Matt was allowed narcotics at only specific doses and frequencies. Sometimes, if the pain abated, he hid his Oxycontin in a wad of gum, spat it onto his lap, and saved the pill for a time when his pain became really intense. Matt had adapted, and he knew that his friend Danny would help him.

Dr. Bloom discovered Matt's system of medication management. He told him, "Matt, I'm trying to be sympathetic. I can imagine what you're going through. I'm your doctor, not your jailer. But this can't go on."

Once they better understood each other, Dr. Bloom slowly tried to increase the narcotic dose and make the schedule less restrictive—without making Matt confused or fatigued.

Visits from Katie Perette, who always brought his favorite food, helped Matt deal with his pain, paralysis, and depression. Cheeseburgers, buffalo wings with extra, extra, extra blue cheese, and chocolate cake were among Matt's favorite foods. Religiously, Katie stopped at Joe's Bar & Grill on her way to visit Matt three times each week, and she then fed him dinner. If especially hungry, Matt would say, "Kate, can you go to McDonald's and get me two number twos? Remember, no onions, only lettuce." Katie would dutifully hop into her car.

⸻

Christopher Reeve soon responded to Matt's e-mail:

I was so sorry to read about your spinal cord injury. There are no words really other than it is simply unfair. However, what struck me most about your letter are your positive attitude and your amazing capacity to adapt to this new way of living. Of course I know firsthand what a struggle it is, but you are already way ahead of the curve. And you are so lucky to have such a wonderful, caring family to be there for you. As you seem to already know, you have a long life ahead of you, and it will be as good a life as you make it.

Unfortunately, though, I cannot answer your question, as Mr. Chase gave you incorrect information. I have not traveled anywhere for stem cell surgery or any other kind of treatment. I did go to Israel this past summer to visit rehabilitation hospitals and to meet with some scientists to investigate the research they are doing. I did not undergo any type of treatment while I was there, and I have not been to Spain or England in years. The only surgery I have had was in Cleveland, Ohio, last spring, where I had a diaphragm pacemaker installed in my chest. I can now breathe off the ventilator for hours at a time. It is still an experimental procedure and not widely available but definitely a step in the right direction.

Matthew, I do believe there will be treatments available soon. Please do not give up hope. Scientists around the world are making very real progress. Many now believe that significant recovery will be possible in the next few years through cell transplants and other methods of promoting regeneration. The fact is that the cord does regenerate. Axons and neurons can sprout and make new connections; dormant pathways that have been spared can be awakened and lead to functional recovery. The key to all this, from the patient's perspective, is exercise. Let the scientists work on biological strategies. Our job is to maintain muscle mass, bone density, skin integrity, good circulation, and a clean, healthy pulmonary system. Don't let anyone tell you what to expect or set limits for your recovery. You need electrical stimulation of all muscle groups and to ride a stationary bike with electrodes at least three times a week. You should stand on a tilt table at least twice a week, and if there's any possibility of getting into a pool for aquatherapy you should grab it. Think in terms of meeting the scientists halfway. You'll feel better, the time will go faster, and you'll have the comfort of knowing that when a new therapy or even a cure comes along, you'll be ready.

I normally tell people living with spinal cord injuries that it's about not giving up. But you clearly haven't. You'll learn new ways to do the things you enjoy, and that your life holds more promise than you can imagine. Please don't forget to dream; nothing is impossible. The main comfort I can offer you—and I believe it is very real—is that it won't be this way forever.

Best to you and your family,

Christopher Reeve

Of Matt's many losses, the insidious and unanticipated one was the continuing loss of many of his friends. How could he have imagined—when his life was severed in two—that his friends were not going to be there forever? He had not learned that friends from high school and college are still friends, but they move away for jobs, get married, and have families. Ellen felt a key factor was that he got hurt at an age when friends are everything. Pat thought that his emotional growth stopped.

One of the biggest problems faced by people with severe disabilities is boredom. Often they are forced to lie in bed or sit in a wheelchair, having few interactions with the outside world. Isolation may lead to medical problems such as depression, lack of compliance with medications and procedures, and eventually life-threatening problems, including infected pressure ulcers. Expanding Matt's sphere of interactions and communication, allowing him to leave the prison of his room, would nourish both body and mind. Many people with tetraplegia can use voice-activated computer systems—to read newspapers, communicate, study, work, and manage finances—but Matt's scarred voice was unreliable and frustrating.

Daydreams allowed him to escape, and he confided in Danny about a recurrent one: *At the line of scrimmage, poised with his knees bent and lower body anchored to the football field, Matt stood his ground and tried to figure out the blocking play. He had to hit hard and get ready to be hit, to ignore the pain in his hips and knees and hands, to plug the gaps on both sides of the center, to fight off the offensive linemen so that his own linebackers could make the play and stop the advance. So Matt got ready, breathed deeply, taking in the smells of the torn-up grass and painful sweat, ready to explode, ready to grab and shove and stop the jump-through of the opposing center*—until the ventilator knocked the wind into him. Matt gasped and returned to his paralyzed and insensate body, far away from the goalposts at the end of his mind.

Pat was aware of boredom's deadly complications and said to Ellen, "He's a bright kid, a very bright kid. It breaks my heart to leave Matt alone in that room. I put on the ballgames. He occupies his time the best he can. I'm not in that situation, so I don't know, but if he tried to get out and go around . . ."

"If he got into a chat room, that would be very good for Matt," Ellen said. "I set up a system he could use with the staff's help. But he was having trouble with his voice. There are only about ten thousand tetraplegic patients like Matt. But I see them chatting online about different things, issues that come up. He needs to feel connected."

"That group's important for another reason," Pat said. "I never want him to give up the dream that he'll walk again. He knows his injury is severe. He's got to have hope, not unrealistic hope, but some hope. Otherwise, he'll just give up."

At the end of 2003, Ellen wrote in her diary:

Twenty-nine months after Matt's injury, he seems to be coming to terms with his life. He talks of the future and the possibility that there may be a cure in his lifetime. He also looks forward to spending time on his computer and visiting with his friends. He no longer blames us for abandoning him. He seems to be at peace with his life. We have set up a foundation in Matt's name to help other victims of spinal cord injury.

Our family goes on. We've returned to work and our everyday responsibilities. At times, we can even laugh. I'm careful not to let myself think too much about Matt's injury and the losses we have suffered. Matt will probably never know the joy of a wife and children. There are places he will never travel to and things he will never be able to do. His brother Mike has lost a buddy; he'll never have nieces and nephews or a sister-in-law. Our family time is relegated to hospital visits with Matt, occasional holiday visits at home, and the annual birthday dinner. In the final analysis, though, we still have our son, and he has accepted his life as it is. We continue to be sustained by our faith, our family, our friends, and hope. And we continue to try to support others who have suffered what we have. Most of all, we try to sustain Matt with hope and love. That has given us a purpose to go on.

Monkeys Playing Video Games

BrainGate, Late 2003

In July 2003, before Cyberkinetics started the first human clinical trial of the BrainGate system, a scouting party visited me at my office in the Southern New England Rehabilitation Center in Providence, Rhode Island. John Donoghue, the chairman of the neuroscience department at Brown University, had formed this corporation to test and market the brain implant and computer system that might allow people with tetraplegia to merely think about moving a cursor and have it move as desired. The Cyberkinetics group had done its homework and knew of my clinical and research experience with spinal cord injuries. Now they wanted my advice on how to conduct the clinical trial. Fascinated by the project, I offered suggestions on selecting and caring for human patients in a research study. For instance, I noted that certain complications associated with spinal cord injuries, such as severe spasticity of the limbs, could adversely affect data collection with the BrainGate.

After the meeting, Cyberkinetics e-mailed me a plan for the first human study of the BrainGate. The company proposed that research subjects would reside at my center for a year after surgical implant of the microelectrode array. We would evaluate the efficacy and safety of the device after the surgery. But my administrative colleagues had a luke-warm response due to the risks of complications and the long-term com-mitment, so I sent back a polite but neutral answer.

With the end of summer, the Cyberkinetics team entered a new sea-son in planning the BrainGate study: patients in the study would stay in their homes or nursing homes. Room, board, nursing, and medical costs were already covered by health insurance, so Cyberkinetics would pay for only the surgery and follow-up care. The research team needed a consultant who was familiar with the medical challenges of patients with complete paralysis. As a result, one day in September 2003, I hosted John Donoghue and Tim Surgenor, the president of Cyberkinetics, in my modest office.

Surgenor, a tall, imposing man with graying hair, was a veteran of the biomedical business, having been an executive vice president for investor relations, quality assurance, and regulatory affairs at another biomedi-cal corporation. Tim chatted about his trip to Providence, but when he started describing the details of the BrainGate study, his words were carefully chosen. He had arrived at his position via a sailing trip in Cape Cod with a former classmate from Harvard Business School, a venture capitalist and investor in Cyberkinetics. Intrigued by the cutting-edge project, impressed with the research team, and flush with more than $9 million, Tim felt it was time to develop a human BrainGate. A leader with his business education and practical experience in medical research was essential for the project.

"What we're doing with spinal cord injury patients is just the begin-ning," he said. "The market for treating spinal cord injuries is not as big as for diabetes or obesity. We're focusing on an area where people get very excited about the science but not the business, so that creates a chal-lenge. But we have to take the first step."

Tim's mission was to develop a quality system, get approval from the Food and Drug Administration (FDA), design the clinical trial, and do the human study. A lot could go wrong between the monkey lab, where the BrainGate was being tested, and the first human BrainGate study, but he felt up to the challenge. After giving this background, Tim turned over the discussion to John.

With his wire-rimmed glasses, white beard, and tweed jacket, John Donoghue looked every part the professor. He fired up his laptop, his expression intense and his speech precise as he compared brain neurons to an orchestra. He wanted to record the activity of multiple neurons simultaneously and hear the full symphony of the brain, which contains a hundred billion nerves and coordinates millions of them for complex behaviors.

John had studied the brain's structure and function for almost three decades. His doctoral thesis on the thalamus, a vital relay station for sensory input to the brain, was completed at Brown University in 1979. (The year of his birth as a full-fledged researcher was also—I later learned—the year that Matt was born.) John went on to publish numerous articles about the brain, but his overarching goal was a form of telekinesis: he wanted to record the signals of individual nerves, predict the brain's intention to move, and convert nerve signals to actual movement—*without* moving the body.

John described breaking through a major barrier in 1996. He implanted an array of microelectrodes, each one millimeter long and thinner than a human hair (see figure 8.1), into the motor cortex in the brain of a sedated and anesthetized monkey. This array—the size of an aspirin pill—was made of silicon, gold, glass, and platinum. A fine bundle of wires conducted the electrical recordings of individual neurons to a pedestal that was mounted on the skull with titanium screws; a fiberoptic cable sent these signals to a computer that stored and processed the electrical messages. The implant site was covered with a cap made of dental acrylic cement.

For three years, John monitored neuronal signals while the monkey made specific hand movements. The animal survived a viral infection in 1998, but it had to be euthanized when the cap came off. As with all scientific research, sometimes the experiment fails utterly, but the process yields valuable knowledge.

These monkey experiments, which occurred over almost a decade,[1] were the necessary groundwork for Matt to receive his brain implant. John and his team improved the microelectrodes, the surgical technique, the software, the computer hardware, and the safety of the system. He explained how his research group tested various methods to cover the surgical opening: first dental acrylic cement, then a titanium mesh, and finally the original bone flap. To pay for this expensive research, John kept applying for and winning federal grants.

John explained that his team at Brown University had designed the experiments as follows:

The microelectrode implant would be placed in the area of the motor cortex that controls the arm. Nerve signals from individual neurons would be recorded, enlarged, and filtered. This would remove electrical noise, or static, and make it easier for the computer to work with the signals. Electrical messages from the brain come through as waves, and these would be converted to digital spikes. Now, the software could correlate the neural signals with specific movements. In this manner, the researchers hoped to use the brain's electrical activity to *predict* the intended movements. With this data base, the final step would be to obtain new recordings, predict the patient's intention, and control a cursor with only thoughts—no hand movements.

With a scientist's pride, John said that the electrical signals of only twenty-four neurons predicted hand activity. The monkey's actual hand movements were compared with those *predicted* by the software. The researchers were delighted to see a close match. They had decoded the neural signals while the monkey was moving its hand—in real time. This breakthrough was presented at the 1999 meeting of the Society for Neuroscience.[2] To describe nerve activity in relation to movement, John used the analogy of a roll for a player piano, which encodes and produces specific patterns of musical notes—a melody—from seemingly random marks on a cylinder.

Continuing his laptop lecture, John stroked his white beard and showed a video that could have been titled "The Dance of the Neurons": a green cursor danced around a red target before hitting it, and when the red dot randomly darted away, the green dot reached it again. This video was of a monkey using only the electrical activity of its neurons to play a video game—without using its hands. A microelectrode array recorded neuronal electrical activity. The BrainGate's software decoded the electrical messages, predicted the desired movement, and used this information to move the cursor in a matter of milliseconds. Because the system quickly substituted for actual hand motion, it might be used by humans with paralysis. This achievement was published in 2002 in the journal *Nature*: "Instant Neural Control of a Movement Signal."[3]

Over almost a decade, forty-one arrays were implanted for up to thirty-three months in eighteen macaque monkeys, proving the Brain-Gate's long-term safety. In general, electrical signals remained normal for extended periods, a reassuring sign that the implant did not damage the brain's cortex. By interpreting the electrical messages of as few as six neurons, the software could predict the direction of hand movement. The monkeys could safely and effectively use the BrainGate as a brain-computer interface.

Clinical usefulness in paralyzed humans would require fast and accurate decoding of neuronal data, which would be used for cursor movement. First, patterns of nerve activity would be *correlated* with a patient's intention to move the cursor in specific directions. These data would be stored and used with later neuronal recordings to *predict* the desired target. Finally, this information would be used to *control* the cursor's movement with thoughts alone.

"Incredible," I said, astonished by the potential of this device for helping people with severe disabilities.

John and Tim looked at each other and smiled; they had won a convert.

We discussed some technical details, then turned to more mundane and practical issues for the first human BrainGate study. In contrast to the six months proposed in the protocol—the detailed plan of the project—I suggested waiting a year after the spinal injury for maximum recovery.

Some complications could also affect the study, and I recommended adding them as exclusionary criteria in our search for the ideal research subject. Severe spasticity once caused a patient of mine to have spasms in which his back arched, his legs shot out in front, and he was thrown out of his wheelchair by his own body. Severe cases of spasticity require an implanted pump to deliver medication directly to the spinal cord. A dreaded complication of tetraplegia is autonomic dysreflexia, in which the systolic blood pressure shoots above 200, sometimes leading to seizures and even fatal brain hemorrhage. Patients with severe chronic pain, requiring high doses of sedating narcotics, might not be able to use the complex BrainGate system. I also suggested the criterion of a "stable psychosocial support system," in view of a higher incidence of depression and illicit drug use in these patients.

Soon after our meeting, Tim wrote that Cyberkinetics "would be very interested in having further discussions regarding the potential of performing the trial as you outlined." I responded enthusiastically in my e-mail back to Tim.

Now I had to determine the official site for the study. I contacted Marilyn Serra, the president of Sargent Rehabilitation Center whose vision included supporting research to help people with disabilities. This center in southern Rhode Island is a vital resource for children with autism, learning disabilities, and various neurologic disorders. In addition, the center offers rehabilitation services for adults with neurologic conditions such as traumatic brain injuries and strokes. Deeply committed to people with disabilities, the Sargent Rehabilitation Center offers free parent workshops, support groups for patients and families, and consultations for schools and community organizations. As I had hoped, Marilyn was interested in hosting the study.

Before sending more information, Tim faxed a confidentiality agreement. Competitors would be curious about the BrainGate technology, which had to be carefully guarded. So I had to maintain the strictest confidence—for seven years—regarding "inventions, products, designs, methods, know-how, systems, processes, software programs, works of authorship, financial records and data, customer lists, projects, plans and proposals."[4]

The agreement was my passport to a rarefied world of biomedical research. Soon I received a FedEx package with drafts of the study protocol and what is called the investigator's brochure. The protocol included the following chapters: "Clinical Protocol," "Benefit/Risk Analysis," "Implant and Follow-up Procedures," "Device Testing," "Monitoring Procedures," "Informed Consent Materials," "Institutional Review Board (IRB) Information," and "Financial Disclosure Form." Vast amounts of data that took almost a decade to collect were summarized in a table and justified the BrainGate's first use in a human being. I learned that the deadline for submitting the research project to the FDA was the end of 2003.

In short, the protocol was the BrainGate Bible. It contained the "Genesis" of the device, "Commandments" for inclusion/exclusion criteria, and "Proverbs" for specific procedures to comply with FDA standards. At the end, the "Prophets" envisioned how the BrainGate would vanquish paralysis and enhance independence.

Throughout the process of helping to develop the final protocol, I kept in mind my ethical responsibilities and medicolegal risks as a consultant and principal investigator for the first human BrainGate study. In spite of our detailed discussion to ensure that patients and families understood the risks, benefits, and alternatives for the BrainGate, there was always a chance that someone could sue after a bad outcome. My contract with Cyberkinetics indemnified me up to a few million dollars, but in this litigious age, there are no guarantees.

I found some reassurance in John Donoghue's monkey research, especially in the long-term safety of implants. Normal movement and behavior in the monkeys suggested that the implant did not damage the brain. Tissue analysis to assess structural and biochemical changes showed only some expected compression of brain tissue. Neurons with normal appearance and electrical activity near the electrodes as well as the absence of neurological symptoms suggested that these effects were not clinically relevant.

Our design of the BrainGate study was based on international "good clinical practice" guidelines for "the conduct, performance, monitoring, auditing, recording, analyses, and reporting of clinical trials that provides assurance that the data and reported results are credible and accurate, and that the rights, integrity, and confidentiality of trial subjects are protected."[5] In the early version of the protocol, candidates for the trial had to be unable to move any muscles below the neck in any reliable or useful way. It seemed reasonable and ethical to allow patients to risk losing movement that was not useful for daily activities. But another consultant asked if people with *any* movement of the arms or legs should be considered for the study. After some debate, we agreed that study subjects could have a maximal grade of one-plus out of five, or only slight arm movement.

One of my closest collaborators at Cyberkinetics was Shawn Wery, the clinical research manager and monitor for the BrainGate study. His impressive title meant that he took care of myriad details while also serving as a taskmaster to consultants like me. In October 2003, he intro-duced himself in an e-mail laden with attached tables for follow-up evalu-ations, lists of safety concerns, and other technical details. He requested detailed comments within three days, so I plunged into the research waters of the BrainGate study with Shawn as my lifeguard. When I met him later on, it turned out that he also looked like a lifeguard—tanned and muscular—and enjoyed the beaches of New England.

I responded to Shawn's thought-provoking queries. In addition to the baseline screening tests in the draft protocol, I requested a preoperative CT scan of the brain. Coexistent brain injuries occur in about 10 percent of people with traumatic spinal cord injuries. It would be disastrous to discover this on the morning of surgery, when an MRI scan would locate the anatomical landmarks; by then, the surgical train would be roar-ing down the tracks. The BrainGate implant would probably not record nerve signals from a damaged brain.

Also, instead of doing lab tests only "as clinically indicated" at the end of the study, I asked that all the preoperative labs be repeated after the explant surgery to remove the microelectrode array, including the CT scan of the brain. This would ensure that no significant changes had occurred.

I also included some rather mundane requests. Because of my expe-rience with electromyography, in which a fine needle is inserted into muscles to diagnose nerve and muscle disorders, I knew the importance of grounded electrical outlets for the patient's safety and for limiting electrical noise.

For the first BrainGate implant in a human, I wanted to be cautious but not so obsessive-compulsive that our search for the perfect sub-ject excluded all candidates. Some criteria, however, were sacred. For instance, no one with seizures within the previous three months should be enrolled, because a seizure could dislodge the implant or lead to bleed-ing in the brain. On the other hand, people with SCI are prone to osteo-porosis, so excluding people with bone disorders seemed too stringent.

Within twenty-four hours of my comments, Shawn followed up with more documents. Because of our tight deadline for the submission to the FDA by the end of the year, I carefully considered Shawn's highlighted questions on Friday evening and sent back the protocol by FedEx the next morning.

||||||||||||||||||||||||||||||||||||||||||

Now part of the Cyberkinetics team, I learned that the company was established in 2001 to develop a commercial product for people who are paralyzed. John and his colleagues felt confident about the safety and function of the implant, so Cyberkinetics evaluated the research from the perspective of the FDA and planned a human trial. This was easier said than funded, so the founders of Cyberkinetics searched out venture-capital funding and added their own seed money. John described a year of expensive and time-consuming studies to thoroughly test the implant's safety. The researchers immersed it in a solution similar to the human body's extracellular fluid to see if it released any harmful materials. John told me that, to ensure electronic safety, the components were shocked with strong electric currents. With limited annual funding from the National Institutes of Health—instead of venture capital—these experiments would have taken ten years instead of one. To achieve the first human BrainGate implant, consultants for Cyberkinetics estimated start-up costs of more than $10 million. This would take three years with private funding, John told me, in contrast to thirty with government support.

Working furiously to meet the FDA submission deadline, Shawn often sent me queries about the study's operational details. When he asked about complications the patient's caregivers should monitor, I suggested watching for urinary infections, pneumonias, and fevers, all of which could infect the implant site. Shawn called for additional information, so I faxed materials from standard texts of neurology and rehabilitation medicine as well as a widely used test called the Mini-Mental State Examination.

As thanks, Shawn sent an e-mail asking for more "standardized tests for these patients." He wrote, "Is it something you can put together?"

With more materials to review and more detailed questions to answer, I felt myself drifting farther and farther from shore, into an ocean of uncertainties. Shawn was still watching as lifeguard, although now with binoculars. In parallel with revising the protocol, he was preparing forms to collect data in a precise and consistent manner. He asked me to help develop these case-report forms (CRFs) for muscle strength, sensory function, reflexes, range of motion at the joints, grading of the spinal cord level, and complications. An infected implant could cause a loss of facial sensation or affect the nerves for speech and swallowing. For a patient with tetraplegia, these losses would be devastating. Worsening of muscle spasms would require medication adjustments, potentially affecting the BrainGate. And, of course, the pedestal site had to be carefully watched for signs of infection. All these potential problems were addressed on the case-report forms in order to monitor patients for changes in their musculoskeletal and neurologic status after the implant.

Psychologically and financially, the FDA submission was an important goal for Cyberkinetics—and the company's investors. In an urgent e-mail, Shawn asked for my draft forms by the next evening. He wrote, "This will allow me to incorporate these into the CRFs." Such was the pace of development for this research study. I was becoming used to serving as a short-order consultant for a start-up biomedical company, so I shrugged, set to work that evening, and faxed off the forms the next day.

Later in October 2003, Shawn sent an e-mail with draft forms, asking for prompt feedback to allow "time to update the CRFs and provide one more review cycle." I suggested using the standardized American Spinal Injury Association form: a schematic human figure with sensory bands related to specific spinal cord levels and tables for muscle strength, reflexes, and range of motion. After we e-mailed and faxed the forms back and forth, Shawn was pleased with the final CRFs.

After this intense phase, I felt gratified to hear from Tim at the end of October: "Thanks to your help, we've made great progress on the clinical plan. . . . We are pressing hard to finalize all of the details and get our filing down to the FDA." Tim's life had been consumed by presentations to investors across the country and strategic planning. He had also recruited Burke Barrett to serve as the vice president of clinical operations. Burke turned out to be an easygoing biomedical executive about my age. Our sons were interested in baseball, so we often chatted about the Red Sox and wondered if they would ever again win a World Series. I soon realized why Burke had been hired by Cyberkinetics. With his remarkable grasp of the operational issues in biomedical research at the medical-device company Cyberonics, he knew all aspects of the process, from protocol design to the final FDA report.

By mid-November, Burke and I often communicated by phone and e-mail. For the institutional review board (IRB) applications, we decided that Gerhard Friehs would go through Rhode Island Hospital, where the surgery would be performed, and my site would use a central IRB that oversaw projects all over the country. An IRB consists of clinicians and laypersons that review and oversee research to ensure the observance of ethical and medical guidelines. This board could decide that the study posed too great a danger, especially in the context of current therapeutic devices for severe paralysis. An IRB application required an official research site, where patients would be evaluated and their records would be stored.

Deeply committed to improving the quality of life for people with disabilities, Marilyn Serra and her board at Sargent Rehabilitation Center wanted to support the study. A major issue for the board was the medicolegal risk of a neurosurgical study, so I explained that almost all of the clinical activity would be conducted at Rhode Island Hospital or the patient's dwelling. In addition, I asked Burke to include the center in my contract's indemnification clause: Cyberkinetics would be financially responsible in case of a bad outcome. Reassured by this legality, the Sargent board approved the BrainGate study, and the rehabilitation center became our research site.

Burke and I approached New England IRB because, unlike other national IRBs, it allowed us to start the lengthy review process without having an FDA approval in hand. My application included a cover letter, submission form, and the BrainGate protocol. This study was risky due to the fact that it involved brain surgery—and that it was the first of its kind in humans—so a lot of detailed information was required on the nine-page IRB form. For example, I had to describe Sargent Rehabilitation Center and explain how we would deal with emergencies. Potential subjects were "vulnerable" because all had a chronic condition; I had to specify measures to prevent their exploitation and preserve their confidentiality. All materials to recruit patients, such as letters and advertisements, had to be approved. I also sent a copy of my medical license, a financial-interest disclosure form stating that Cyberkinetics paid me a consultant fee, and the investigator's brochure.

Probably because of the impressive monkey research at Brown University, New England IRB had only a few suggestions for Burke after our preliminary inquiries. We stated that patients would not pay for tests and procedures. I asked Marilyn Serra, as president of Sargent Rehabilitation

Center, to address any concerns from patients and families. Another required revision was a more complete listing of possible complications.

Of all the study documents that I helped develop and all the forms I completed, the most daunting was US FDA Form 1572. It had a section titled "Commitments" that required me to follow the protocol, to obtain proper informed consent, to report any adverse events to Cyberkinetics, and to keep adequate and accurate records. A warning followed these and other commitments: "A willfully false statement is a criminal offense." In human research, a lapse in ethical and clinical standards can lead to the loss of research privileges, medical licensure, and even freedom, in case of a criminal conviction.

Completing these forms and embarking on this project was not to be taken lightly.

9

A Walk Around the Lake

Linda Holmes

Linda Holmes and her husband, Barry, were at home on February 27, 2010, when there was a power outage because of heavy snow. While waiting for the electricity to return, they made coffee with a portable stove and enjoyed a leisurely Saturday morning.

After coffee, Linda washed the dishes and turned to Barry to comment on the weather. He looked at her and said, "Are you OK? Your face is drooping."

"I feel perfectly fine," Linda said.

After twenty-six years of marriage, Barry knew every feature of his wife's face and worried about her changed appearance. It occurred to him that the portable stove might have caused carbon monoxide poisoning and affected Linda's brain. A former emergency medical technician, he realized the urgency of the situation and called for an ambulance. Within twenty minutes, Linda received high-flow oxygen, and the treatment fully cured her facial droop. They felt thankful that their exposure to carbon monoxide had been limited, but to be on the safe side, Linda and Barry decided to leave their house and visit his parents.

"Half an hour later, the symptoms came back," Linda told me in May 2011 when we discussed her experience. "I dropped my gloves and keys, and then I couldn't use my left hand. My leg was dragging by the time we got to the hospital at noon." After a CT scan and an MRI, forty-eight-year-old Linda was diagnosed with a stroke.

Four days later, when her condition had stabilized, Linda was sent to Whittier Rehabilitation Hospital in Massachusetts. Her daughter, Rebecca, worked in the billing office and was able to speed up the transfer. Linda stayed there for the next five weeks. Her stroke was so severe, with complete paralysis of her left arm and leg, that she was dependent on her caregivers for almost all her basic activities. It was a week before her balance improved enough for her daughters to give her a shower.

"It was like heaven, having that first shower," she reminisced and laughed. "There I was, half paralyzed, but I worried about being clean."

Linda needed a lot of help with her basic mobility. Moving from the bed to her wheelchair was difficult because her left ankle was unstable and weak due to the stroke as well as an old ankle injury.

"I couldn't move my ankle from side to side," she told me. "And I tried wiggling my toes, and it was like, 'Oh my God!'"

Linda shook her head, still in disbelief at her paralysis.

A week after her admission, she started moving her toes and foot, so her physical therapist supported her lower limb and gave her a hemi-walker. She had to use this four-legged aluminum device on her right side because her paralyzed left arm could not manage a regular walker. (Anna Iacono also used this equipment.) But Linda was glad to be upright, in any way possible, and pushed her half-paralyzed body to walk forty feet. Along the way, her physical therapist kept offering to stop so that Linda could fall back into her wheelchair, but she marched on.

Two weeks after her admission, Linda's leg and her balance remained impaired. Shifting her weight was difficult, and she mistakenly believed that she was putting pressure on her left foot.

"But I wasn't," Linda told me. "And I wasn't bending my knee, but I couldn't really tell." She looked at her left leg, ruefully acknowledging the effects of the stroke on her sensory function. "You take so many things for granted."

In spite of her lack of awareness, however, Linda was able to advance to a cane that had four prongs. She did her best to improve her mobility, but it was a struggle—with her weakness, sensory problems, and leg tightness.

"The hardest thing was how to forget everything I'd learned about walking and redo it," Linda said. "I knew how to walk, I knew how it felt to walk, but it wasn't working."

"The tendency is to keep your knee straight," I said, trying to explain the process I've seen in many patients with strokes. "That helps you stand up. But when you learn to walk again, you've got to bend that knee or you won't be able to bring your leg forward."

"Yes," Linda said. "I tried swinging my leg around and lifting up my hip."

"That was hard work, wasn't it?"

"Yes, it sure was."

Linda had increased extensor tone that straightened her knee. This is the body's way of adapting to the stroke. It is far better to stand up and move for short distances instead of being bedridden or limited to a wheelchair. But walking is difficult because the knee does not bend when the leg swings forward. So Linda leaned toward her right side while lifting up her hip on the left. She also swung her left leg out to the side in order to bring it forward. Both these compensatory patterns increase the energy cost of walking.

While watching Linda during physical therapy, her neurologist, Dr. Joan Breen, thought she might benefit from Tibion's bionic leg. At a stroke conference in San Francisco, Dr. Breen had discussed the robotic leg with physical therapists who had used it for treatments and research studies. The neurologist was impressed with the potential of the robotic leg; she believed in neural plasticity, which requires many repetitions to help rewire and heal the brain after a stroke. It was also crucial for Dr. Breen that patients initiated the movements, taking an active role in their rehabilitation. Anna Iacono and Pat Wines were able to control their knees well enough to stand, but Linda's knee tended to buckle. The bionic leg would help her stay upright.

Based on the concept of neural plasticity, the Tibion leg is a robotic brace that is strapped around the patient's thigh and leg. A shoe insert senses how much pressure is exerted by the leg when a patient tries to walk or sit down or climb a step. At the knee, another sensor measures the changes in the angle of the joint. The bionic leg's computer determines the patient's action and intended action. For instance, if the angle of the knee changes from 90 (bent) toward 0 degrees (straight), the computer predicts that the patient wants to stand up; for sitting down, the converse is true. Motors in the bionic leg assist the patient's desired movement. When the patient descends a staircase, for example, there is increased force at the foot pressure sensor, and the knee is bent. The motor also acts as a brake to allow gradual bending of the knee and prevent a fall.

Imagine a patient with a stroke trying to walk. As the leg swings forward, there is no weight on the foot pressure sensor, and the knee joint moves freely. When the heel is planted on the ground and the knee is straight, the foot-pressure and knee-angle sensors send messages to the computer; a motor in the robotic leg helps the patient to remain standing.

Walking is a rhythmic and repetitive process, so the bionic leg's computer can anticipate the intended movements. Based on these data, the computer activates the motors in the robotic brace to support and enhance whatever strength is left in the muscles surrounding the knee. By evaluating the patient's strength and balance, the physical therapist determines the support required from the bionic leg and programs this into the computer for various activities. Most important, the patient always remains in control by initiating the action. With improvements in the patient's strength, balance, and mobility, the physical therapist decreases the support provided by the bionic leg.

Eventually this combination of physical and robotic therapy may allow the patient to exert a force similar to that of the unaffected lower limb. The physical therapist also tries to reduce the patient's reliance on abnormal biomechanics such as hip hiking to enable the foot to clear the ground. With the device, therapists can retrain patients on a variety of surfaces, starting with hospital linoleum and progressing to stairs and outdoor terrain. Ideally the patient will be weaned from walking aids such as canes and walkers. Near the end of the therapy program, the patient should be able to walk with no devices except the robotic leg. Through repeated use of the bionic leg as well as physical therapy for balance and strengthening, the patient strives to walk independently—eventually without the Tibion leg.

Preliminary research has shown that the beneficial effects of Tibion's bionic leg are maintained even when the brace is removed. This is probably due to strengthening of the limb and rewiring of the brain, allowing patients to improve their mobility even without the robotic device.

At first, Dr. Breen prescribed the electronic brace for more advanced patients, but she also used it to prepare patients for walking.

"It has helped me get people out of wheelchairs so they get maximal recovery," she told me as we discussed Linda's rehabilitation. "Even after one session, I've seen a better gait pattern. And the bionic leg can be used at any stage of recovery."

She felt that the weight of the device could be a factor, but Tibion has developed a lighter model that weighs less than ten pounds.

"Patients can tell that they're walking better, so they want to use it," Dr. Breen said. She also felt that the device reduced the physical work of the therapist, who could now focus on improving the patient's gait pattern.

Linda saw the device about two weeks after her admission. Her initial reaction was that it looked strange (see figure 9.1).

"I had to trust that it would hold me up," she said.

Her one-year-old granddaughter, Jaidalynn, was afraid of the device at first and said, "Oh no, Nana!" But when she saw Linda walking and heard the toylike noises, she said, "OK, go, Nana."

Linda at once realized that the device helped her physical therapy, for it allowed her to shift her weight onto her affected leg. The robotic leg also stabilized her knee and made her less afraid of falling.

"In the beginning," she told me, "it corrected the way I was walking. If I didn't move my knee the right way, it made these whirring and beeping sounds. It talked to me all the time when I wasn't stepping right. But as I got better, it didn't talk much."

Auditory feedback helps patients to shift weight onto the weak leg and work with the motors of the robotic leg. With the Tibion leg, Linda made good progress with her mobility and walked 172 feet. It was the greatest distance she had walked in one session after the stroke.

"We used it mainly to gain balance and strength. It helped a lot going up and down stairs."

As her strength improved, the physical therapist decreased the support from the device.

"Did you feel tired after using it?" I asked her, wondering about the weight of the device.

"After a while I did feel a little tired," Linda said. "But when I took it off, I could tell my walking was better."

"I'm sure it strengthened your muscles," I said. Having heard only the benefits of the robotic leg, I asked Linda, "Did you have any problems with the device?"

"Every once in a while the foot sensor would give out," Linda replied.

"Tibion is coming out with a new version," I said. "I bet they'll fix that problem. How're you doing with the ankle?"

"It still feels stiff, and I wear a rocker-bottom shoe to help me walk. But I don't have much of a foot drop; I can put down my heel and lift up my toes. That's a big accomplishment."

"Yes," I said. "It is. That can be very difficult after a stroke. Some of my patients never achieve that."

A few weeks after using the bionic leg, Linda graduated to a straight cane. As her mobility improved, she started therapy in the hospital's heated pool. Every Thursday she worked on her balance, strength, and endurance. She continued using her plastic brace because her ankle was weak from the stroke and unstable due to her old injury.

"I used the bionic leg to help me with standing and sitting. I used to lean to one side, which I didn't realize until I saw myself in the mirror. The therapist had me put a ball between my knees to correct the position of my hips. I also did a lot of sit-to-stand exercises to strengthen the core muscles."

"How often were you using the device after leaving the hospital?"

"Once every other week," Linda said. "I used it about an hour at a time."

"Was it helpful in retraining the way you walked?"

"Yes, it was," Linda said.

As her mobility improved, Linda discarded her plastic ankle brace in August 2010, six months after her stroke. That month she also became fully independent with bathing and dressing. A few months later, in January 2011, she was able to cast away the cane.

"That's great progress," I said.

"I pushed it," Linda said.

She used the robotic brace until March 2011, when she needed only supervision for short walks. Linda's strength had improved to the point where she could lift up her thigh and extend her knee against resistance, but bending her knee remained a problem.

Why was the bionic leg so successful for Linda? At a Tibion presentation, I used the device and found that it worked in concert with my own effort as I moved around. When I moved my leg, the robotic brace assisted my muscle actions, especially with staying upright and with stairs. For instance, when I pretended to fall by letting my knee buckle, the brace sensed my risk of falling and straightened my knee. My experience with the bionic leg made me realize its potential for helping patients.

Often the simple act of rising from a sitting position can be insurmountable for patients with strokes, and they may restrict their mobility due to a fear of falling. This anxiety is unfortunately based in reality, as almost 40 percent of patients have a serious fall during the first year after a stroke.[1] With the robotic leg, patients feel more confident about standing up and moving around. Consequently, they are more mobile and will perform many more repetitions of the complex movements involved in standing up and walking. They are better able to negotiate real-life challenges, such as curves, stairs, and uneven terrain; these activities assist brain healing through neural plasticity.

Theoretically, as the patient increases the repetitions of various movements, the damaged brain is rewired. In fact, patients have used the bionic leg and seen improvements years after a stroke. For example, a fifty-seven-year-old man had a stroke nine years before using the bionic leg. Because of weakness and spasticity, his mobility was limited. After eight hours of therapy with the Tibion leg over a period of four weeks, he improved his balance, endurance, and walking speed. A month after finishing this therapy, he retained these gains. Another patient used the device five years after his stroke and improved his walking speed, endurance, and mobility with shopping and other activities.[2]

Linda's knee required strength as well as support, so the Tibion leg was an ideal device. It helped her to use her stroke-weakened knee muscles, which made them stronger and eventually supported her knee without the electronic leg.

Linda still needs a red-and-black elastic support for her ankle. "I call it my boxing glove; I slip it on and lace it up. I have to walk slowly so the ankle won't twist. When I go for aquatic therapy, I have to be really careful."

"How often did you do pool therapy?"

"Every other week," Linda said.

"Was it the same week as the Tibion leg?"

"No, we alternated so that I got therapy every week."

"Sounds like you had a well-designed program."

Linda agreed and said that, when her balance got better, she started the treadmill.

"I'm not a competitive person, but I compete with myself. If I walked 2.5 miles per hour, I'd push it to 2.6. I kept challenging myself. If it was too much, I'd back off. If I was told I couldn't do something, I had to prove them wrong. That's the way it is with the stroke—I just can't lie down and roll over. I had to push until I'd gone as far as I could go. Now I've accepted that it is what it is. I'll learn to live with it. But as time goes on, I'll try something new and build on that."

"Can you give me an example?"

"Well, I had a hard time opening jars," Linda said. "Then I noticed my strength was coming back, so I tested myself and opened a spaghetti-sauce jar. I keep revisiting things like that so I can see my progress."

"That's a good approach—using your progress as a benchmark, like an athlete."

"Yes, but I never thought of myself as an athlete," Linda said.

"Stroke rehab is a very difficult event," I said. "It takes more strength than Olympic weightlifting and more endurance than a marathon. And it goes on for the rest of your life."

The stroke had a profound impact on many other aspects of her existence. Linda had been a shipping and receiving manager for a sheet-metal company but had to go on disability. Then she lost her health insurance in August 2010.

"We couldn't afford it because it cost eleven hundred dollars a month," she explained.

"What about disability?"

"I was only getting eighteen hundred for that, and it costs twenty-five hundred to run the house, groceries, utilities, car, and all the basic necessities. I had to give up some medications. We made too much money to get help but not enough to pay for all the meds."

"Which ones?" I said, hoping she wouldn't name one of the medications prescribed to prevent a stroke.

"Aggrenox."

I felt sad that Linda couldn't use this blood thinner to prevent another stroke, as her risk was nine times that of an average person.[3]

"Are you on aspirin?" I asked, hoping that she at least took this medication to prevent a stroke. (According to drugstore.com, a year's supply of Aggrenox is about $2,600, while aspirin is about seven dollars.)

"Yes," Linda said.

I was relieved to hear this, but I thought I'd better ask about other risk factors.

"Do you smoke?"

"I started smoking as a teenager, but I quit the day of the stroke. I haven't smoked since."

"Great! Do you drink?"

"No." Linda shook her head and said, "I don't do any of the bad stuff anymore."

"Well, keep it up," I said. "You've got many good years in you."

I asked her to describe a typical day.

"I have breakfast and coffee, and then I go online."

For a course on medical coding and billing, she had studied five hours each day and recently got her certification with a perfect grade point average.

"I wanted to work for the hospital that helped my recovery," she said. "In August I'm starting as a patient accounts representative and medical biller." On occasion, she also takes care of Jaidalynn, who no longer worries about her Nana. They enjoy each other's company, and Linda believes that her granddaughter has been one of her best therapists for her stroke. Twice a week she goes to Whittier Rehabilitation Hospital for therapies, including the treadmill and pulley exercises to stretch her arms.

When I asked if the pulleys had helped, she said, "I can shave my armpits now. I can't shampoo my hair with the left hand, but I'm working on it. I do the cooking and cleaning, but my husband won't let me cut the food."

"That's wise," I said.

"Every once in a while, he catches me doing that." Linda grinned at this confession.

"What about other housework?"

"Barry does the laundry, because it's in the basement. He's become a good laundry guy." Linda laughed. "Sometimes I put the basket on the stairs and go down backward with it."

"I don't recommend that," I said. "That may not be safe either."

"OK, I'll be careful." Linda paused to consider my unsolicited advice. "I'm still getting used to this lifestyle. I won't get better overnight."

"No," I said. "But you're doing great."

"When the stroke happened in February, I thought I'd be better by September. Now I know it's more serious than I thought. I have my moments, but my family's supportive. And they've been with me since day one, especially my husband. If it wasn't for them, I couldn't have gotten through this."

"Yes, that makes a big difference," I said. "Do you have a home exercise program?"

"I do balancing exercises, and I use the Wii at home."

"Really?" I said. "Some of my patients enjoy that. I played tennis with the Wii and could tell it's good for rehab."

"I like it for bowling and yoga," Linda said. "That helps my balance a lot."

"You've taken a very creative approach with your own rehab. At first you used the brace and the cane. Then you used the Tibion leg. Now you're using the Wii. What about other leisure activities?"

"Barry and I go camping in our thirty-foot trailer. We like the campgrounds at Barrington Shores in New Hampshire."

I imagined Linda walking along the lake at Barrington Shores—the warmth of the sandy beach on her bare feet, a kayak's paddles splashing in the deep blue water, the sunlight shining down on her through a grove of birch and maples—as her lungs were cleansed by the lake air.

Navigating a Study Through Two
Research Review Boards

BrainGate and Cyberkinetics,
December 2003 to Early 2004

As the first human study of the BrainGate prepared to set sail, Cyberkinetics scheduled an evening meeting in December 2003 at its offices in Foxboro, Massachusetts. Snow was piled high along the road, and I almost drove past the entrance to the large brick building that housed numerous corporations. This meeting was mainly for principal investigators (PIs), the clinicians who would recruit patients and conduct the study. At the conference table were John Donoghue; Steve Williams, chairman of rehabilitation medicine at Boston University; Mijail Serruya, the lead author of the *Nature* article on the BrainGate in monkeys; and Leigh Hochberg, a neurologist at Brown University and Massachusetts General Hospital. The Cyberkinetics representatives included Tim Surgenor, Burke Barrett, and Abe Caplan, the clinical engineer who would operate the BrainGate computer system. Gerhard Friehs, the neurosurgeon for the monkey experiments, was there in a dual role as a clinician and a cofounder of Cyberkinetics.

Our research group discussed various details of the protocol, and then Burke led a mock question-and-answer session, excerpted below, to prepare for questioning by research boards:

Why does the implant have to be removed after a year?
This was the protocol approved by the FDA, and we would offer an extension study for implanting a more advanced version of the BrainGate.

What if a patient decides to keep the implant?
It would then be the PI's responsibility to follow up and ensure that there were no safety issues.

Should we exclude patients with a resistant form of staphylococcus aureus (MRSA)?
After a lively debate, we decided not to include these patients because this was the first human study. As a principal investigator, I felt uncomfortable about liberalizing the criteria for enrollment and had already suggested diabetes as an exclusionary criterion because of its higher risk of infections and poor wound healing.

We also discussed some important clinical questions:

When should we become concerned about fevers?
Because patients with spinal cord injuries are sensitive to environmental temperatures, instead of an absolute threshold I suggested watching for a change from the patient's baseline. During my residency a patient with tetraplegia had a high fever, and I was concerned that a pressure ulcer on his back had become infected. Then I learned that he used a waterbed—filled with warm water. We changed the water, and his fever subsided. This was the easiest and most successful "cure" of my entire career—with no side effects.

How would we approach neurosurgical complications?
As the neurosurgeon, Gerhard Friehs was understandably concerned about infections or bleeding in the brain, which could be fatal. But certain complications were expected, such as inflammation and drainage at the surgical site. We discussed strategies for minor and major complications.

Then there were some philosophical questions, such as the following:

What was the highest level of spinal function that should be allowed for patients in the study?
All along, I had wanted to enroll only people with severe paralysis, who would not risk the loss of movement and could benefit from the BrainGate. [People who can only raise their arms and the wrists against gravity—intact spinal cord at the sixth cervical level, or C-6—are surprisingly independent with wheelchairs, dressing, grooming, and bathing. When the wrist is lifted up, the flexor tendons of the hand tighten, resulting in a limited grasp for certain functions. Some patients can use splints to maximize this phenomenon, called tenodesis.] We compromised on the criterion of neurological function at C-5 and above. At this level, people can raise their arms against gravity and bend their elbows, but not much else. At C-4, as with Matt, there is no movement at all, except the head, neck, and a shoulder shrug.

For dinner, we filed into the office kitchen for sandwiches, potato chips, and soda. It was getting late, so we resumed our discussion.

Who would be interested in the study?
We planned to search for people who were altruistic as well as pioneering, who saw this as a landmark study for disabled people and would go where no one had gone before.

How should we approach the discussion for informed consent?
We agreed to be candid about the risks, for ethical and medicolegal reasons. [Cyberkinetics indemnified the investigators, but the damages could go well beyond the amount listed in the contract. Although the risks of the BrainGate study were discussed briefly, they were always present in my mind. I became uncomfortably conscious of the implications for my career. If all went well, Sargent Rehabilitation Center and I would get favorable publicity. Conversely, if there were major complications or a death, my name could be prominent in newspaper and TV stories in Rhode Island and possibly nationwide.]

How should we handle media relations for this high-profile study?
It was likely that the media would want early results in order to get breaking news. The investigator would work closely with Kari Watson of MacDougall Biomedical Communications, the media consultant for Cyberkinetics.

What if our first patient or family member approached the media at an early stage? What if a journalist who knew about the study located the first BrainGate patient and tried to scoop everyone else?
Our media consultant would advise us how to best handle such an event. We decided to ask patients and families not to discuss details of the study. After all, initial results could be either incredibly successful or terribly disastrous but not scientifically meaningful with only a few patients.

When I got home late that night, Nita and Jacob were already asleep. My wife, Giselle, woke up when I arrived, and we talked about the research project. As a physician and a wife, Giselle had concerns about the invasiveness of the study and the risks to the patient—and to my career. So I told her only about the positive and reassuring features of the meeting and tried to help her get back to sleep.

But sleep did not come easily to me. I lay awake, my cognitive neurons firing more actively than usual, as I envisioned how our team would navigate the reefs, shoals, and rocks upon which our research study could founder. I considered all the barriers we had yet to cross: the review board at Rhode Island Hospital, the New England review board, the FDA, protocol revisions for issues raised during reviews, and, finally, finding patients willing to participate. Then I started thinking about the risks of the experimental neurosurgery that would be done for the first time in human beings.

Shortly after the meeting at Cyberkinetics, we gave a preliminary presentation to the institutional review board (IRB) at Rhode Island Hospital, a ten-story structure with a semicircular driveway that curved around a center island with a large maple tree. Near the entrance, flanked by large earthen flowerpots, stone steps led to a grassy area with rhododendron bushes and stone benches, an oasis in the middle of disease and death.

I walked through the revolving doors into the main lobby, where the centerpiece was a large marble table with a vase of roses and lilies. Near the service elevators where a nurse waited with a patient on a stretcher, I glanced through the tinted windows at juniper bushes and tall Norwegian pines. Then I entered a sunlit, glassed corridor with a display case of figurines created by the pediatric patients or donated to the hospital: dolphins, dinosaurs, model cars, and horses.

The small door leading to the conference room did not prepare me for the large inlaid table that stretched almost to the other end of the space, where a white screen for presentations hung from the wall. Nor did I feel prepared for the group of around twenty clinicians, administrators, and laypersons in white coats and dark suits.

In addition to these impressive people, from the distance of history, two tragic and ethereal representatives keep watch over every IRB. They are present with us in spirit, historical figures who suffered from violations of ethics and human rights. Clinicians must never forget them. One is a young woman who endured brutal Nazi medical experiments at the Ravensbrück concentration camp during World War II. She testified at the Nuremberg trials as follows:

> In the hospital we were put to bed and the hospital room in which we stayed was locked. . . . Then a German nurse arrived and gave me an injection in my leg. . . . Then I lost my consciousness . . . and I felt severe pain in my leg. . . . In the morning I noticed that my leg was in a cast from the ankle up to the knee and I felt a very strong pain in this leg and the high temperature. I noticed also that my leg was swollen from the toes up to the groin. The pain was increasing and the temperature, too, and the next day I noticed that some liquid was flowing from my leg. . . . The bandage was removed, and that was the first time I saw my leg. The incision went so deep that I could see the bone. . . . The pus was draining from my leg; the leg was swollen up and I could not walk.[1]

In all, eighty-five witnesses testified about their suffering. In addition, the Nuremberg judges reviewed about fifteen hundred documents during the trials. In 1947, the judges found sixteen Nazi "doctors" guilty and sentenced seven to death for their heinous experiments. Thousands of Jews and prisoners of war, including children, were subjected to sterilization, traumatic brain injuries, bacterial infections, and various drugs. Enshrined in the Nuremberg Code are three fundamental principles: (1) voluntary consent, (2) minimal risks, and (3) the expectation that the results will help other patients and society at large. Consent should be obtained by informing prospective subjects about the study goals as well as the risks, potential benefits, and alternatives; they must also be given details about the device, drug, or procedure.

The other permanent guardian of an IRB's ethics and conscience is an old African American man who festered with syphilis as a result of inhumane medical "treatment." Even though penicillin had long been available to treat him, the white "doctors" in Tuskegee closely watched him and others, from 1942 until 1972, to learn how the disease slowly destroys the body:

> His aorta is inflamed and sometimes the pain claws at his heart. When the blood backs up into his lungs, he becomes short of breath after picking only a few ears of corn from his garden. Why do his headaches and his memory and his vision keep getting worse, even though the nice doctors have treated him for years? At first his feet burned, as though soaked in acid, but he can't even feel them anymore. Is that why he falls down on uneven dirt roads? And why are his joints being eaten away by arthritis? And why do the doctors keep doing tests on him? Questions enter his mind like flashes of sunlight piercing the fog of his syphilitic dementia.[2]

The Belmont Report (1979), a landmark document on ethical principles and guidelines for human research, cited the Nazi inhumanities as well as the notorious Tuskegee study in developing its human research standards.[3] This report addresses the boundaries between clinical practice and research; basic ethical principles (respect, beneficence, justice); informed consent; assessment of risks and benefits; and the selection of subjects.

Keenly aware of the ethical issues in human research, Tim Surgenor first gave the group an overview of the BrainGate project. When John

Donoghue presented the monkey videos, I could sense the astonishment of the IRB members. Gerhard and I discussed the clinical aspects, including details of the surgery, recruitment of subjects, and, most important, safety monitoring. As a relative newcomer to the project, I felt nervous in front of the board but was reassured by a few nods of approval from John.

After the meeting, we felt good about the presentation and discussion, but we didn't count on the IRB's blessing. Our research team worried that the board would consider the procedure too risky and too experimental.

All of us were relieved when the IRB did not send a summary rejection of our study. To address the board's concerns about patient safety and education, Cyberkinetics produced a guide for patients and caregivers that explained the technology and necessary precautions. We also expanded the list of potential adverse outcomes, which already included dizziness, fever, bleeding, infection, and death. A checklist of nine clinical items—such as temperature and surgical-site condition—to assess the patient's status would help prevent these events.

After this preliminary review, the Rhode Island Hospital IRB wanted a more focused discussion of the BrainGate study. From the parking lot I walked past the building for ambulatory patient care, a tall and modern gray structure near a large sycamore tree. Then I went through the hospital's sliding doors and past a waiting room and registration area. In the large and busy cafeteria, Gerhard, Burke, and I waited near the conference room for the IRB meeting. We didn't know what to expect. An IRB in another state had rejected the BrainGate study—they had deemed it too risky. My anxiety was at a peak. To prepare for the IRB's questions, we quizzed one another about the safety of the device, ethical issues, and clinical procedures.

Gerhard and I—the study clinicians—went into the meeting while Burke waited outside. I was relieved that none of the technical aspects of the study were challenged and no one raised any major issues that would prevent approval.

Near the end, however, a member asked, "How do you plan to address the fact that this is a vulnerable population?"

Another person chimed in, "And how will you educate patients about alternatives to the BrainGate?"

Gerhard glanced at me, passing on these questions.

After a moment to consider these points, I said, "It's true this is a physically—and emotionally—vulnerable population. But I've found that people with disabilities are keenly aware of the latest research and treatment options. They're invariably better educated about their condition than the average patient. For the informed consent, we'll have an impartial witness as well as the family or caregiver. They'll help protect the rights of potential subjects."

I paused after this short speech, then added, "As a rehabilitation physician who advises patients about medical equipment, I'll discuss alternative technologies."

The board members who had raised the questions nodded, and one said, "Fair enough."

Gerhard and I were excused, and the IRB coordinator said that she would contact us. Burke had finished his third cup of coffee and was glad to hear there were no major problems. But we still could not count on the IRB's approval.

Cyberkinetics allowed the research team a short winter break after the FDA and IRB submissions. Then Burke asked me for a report on assistive technology. In the National Library of Medicine database, I found about a hundred research articles, including one as early as 1985 for a primitive voice-recognition computer.[4] As part of my research, I also spoke to a person with tetraplegia who was quite pleased with her voice-activated system, which controlled lights, room temperature, her phone, and her apartment door.

During dinner with Tim and Burke at a restaurant near Brown University, between bites of linguine in a seafood sauce, I discussed some of the BrainGate's competitors. An environmental-control unit called the CAT-Solo was controlled by movements of the head, mouth, or tongue. A speech device called the QPointer Voice allowed a person with paralysis to create and edit text, manage e-mail, and surf the Internet. The SmartNav 4:AT had a computer screen with a virtual keyboard that was controlled by head movements, so it was slower than voice-activated systems; by dwelling on an icon, the user could left click, right click, and double click. Another device, the EyeTech computer, used a camera that focused on one eye to determine the direction of the gaze, and the cursor was placed at the gaze point. A mouse click required a slow eye blink.

At the end of January 2004, Dr. Celia Witten from the FDA's Division of General, Restorative, and Neurological Devices wrote that our application was "conditionally approved" and we could begin our "investigation, using the revised informed consent document." We had to be clear that "the risk of infection is serious and to provide a list of the possible sequelae . . . e.g., brain edema, meningitis, death."

A few of the many possible adverse events listed in the protocol were not included in the informed consent form (ICF), so this addition was required in language that was "understandable to the layperson." Dr. Witten wanted us to emphasize "that the BrainGate system is an *investigational* assistive device" that would "*potentially* allow physically impaired persons to control a cursor on a computer screen by using their thoughts."

Our research team addressed nineteen detailed comments, such as exactly how we would use the MRI scan to locate the implant site in the primary motor cortex. We also had to exclude people "being actively treated for a serious medical, neurological or psychiatric disease or disorder." Although daunting, all of the FDA comments were quite clear, and we responded within the deadline of forty-five days.

|||||||||||||||||||||||||||||||||||||||||

Comments from the Rhode Island Hospital IRB arrived in early February, with input from the nurse manager for the neurosurgery intensive care unit. As expected, safety was a primary concern.

One IRB member asked about the consequences of damage to the motor cortex (for instance, bleeding caused by the implant). This would not be clinically evident in paralyzed patients, but a large stroke could be fatal.

There was a question about "the direct benefit to participants," and Gerhard explained how the BrainGate could improve their quality of life.

Another IRB member noted that a patient could refuse the surgery to remove the device after a year, and Gerhard responded that I would follow patients as long as the device was implanted. (The study could turn out to be a long-term commitment.)

The antiseptic protocol was questioned, and Gerhard stated that it was standard procedure for craniotomies.

Finally, the board wanted the consent form to "emphasize that the device has not been tested in humans and it is possible that unforeseen adverse events may arise." Pending FDA approval and our written response, the IRB deferred its verdict.

Around the same time, Burke sent me the investigator agreement, which required me to sign off on various statements:

By training and experience, I am qualified to serve as a Principal Investigator (PI) for the study. If my clinical care deviates from the Protocol, I will notify Cyberkinetics. I will provide the Protocol to all health care personnel involved in the study. I will conduct the study in accordance with the Investigator Agreement, the Investigational Plan, FDA regulations, and IRB requirements. I will obtain informed consent from research subjects. I will respond to requests for patient information in a timely manner. If there are any problems with the BrainGate, I will immediately inform Cyberkinetics. I will allow the FDA to audit my study site and review study documentation. I agree to indemnify, defend and hold Cyberkinetics harmless from any failure on my part to adhere to the Investigator Agreement.

I considered all the contractual details, to ensure that I could comply with every requirement, and signed.

New England IRB concluded that our study was "satisfactory in terms of safety and protection of the rights of the research subjects." But it withheld approval pending the resolution of specific concerns: What if the device stopped working? What about device expenses for late complications? What if the subject did not wish to have the implant removed?

Our informed consent form was extensively marked up, though with fairly minor and reasonable revisions. Finally, because this was a "high-risk study," New England IRB would reevaluate it after the enrollment of two subjects.

Burke, Shawn, and I had many phone and e-mail discussions before we addressed the IRB's questions in a letter dated February 12, 2004. In the consent form, we listed the consequences of a brain infection, including swelling, coma, meningitis, and death.

New England IRB had commented on some "medical jargon," so we simplified the language. We clarified that the subject's caregiver would "take the temperature, clean around the pedestal site, and replace the Biopatch," which is an antiseptic sponge placed at the surgical site.

A few days later, Shawn sent me the final documents, including the latest version of the protocol. After reviewing our responses to New England IRB one last time, I signed off on the forms and sent them back by FedEx.

When I spoke with Shawn, he said, "Great! It's exciting that we're so close to beginning the study."

Another challenge was how to publicize the study and recruit patients. Judith Hammerlind Carlson, the president of an assistive-technology company called TechACCESS, was intrigued by the notion of rechanneling brain signals to a computer. At a meeting with Tim, Burke, and me, she kindly offered to describe the BrainGate in her newsletter for people with severe disabilities.

Our publicity net was cast wide. At Shepherd Center in Atlanta, the vice president of research and technology, Dr. Michael Jones, welcomed a presentation. It so happened that Nita and Jacob were on a school break in mid-February, so we made it a family trip. Burke saw the potential for networking and decided to meet me there.

In Shepherd Center's large conference room, where a small group of clinicians and bioengineers had gathered, I gave a presentation with slides and a video of the monkeys using the BrainGate, followed by the study design for the first human implant. I explained that the microelectrode implant would record nerve signals from the brain of a paralyzed patient who imagined moving a computer cursor. These signals would be enlarged (amplified), cleaned up by removing noise (filtered), converted to a digital format (in contrast to analog waveforms), correlated with movement, and stored. Then the patient would try to move the

cursor by thought alone. Electrical patterns related to the desired movement would be interpreted by the BrainGate system using the stored
data, and the cursor would move.

Having given permission for the audience to interrupt me, I was surprised when no one spoke up, but at the end we had a thought-provoking
discussion. Afterward, Michael Jones gave us a tour of the impressive
facility where Matt had received his rehabilitation (although Matt and I
had yet to hear of each other).

To further publicize our research, Burke suggested that we submit
a poster for the annual meeting of the American Academy of Physical
Medicine and Rehabilitation (AAPM&R). John Donoghue had new monkey data, so we would not be recycling old material. In the abstract, we
noted that eight implants were placed in six macaques for an average of
ten months per implant. There were only minor complications, and all
three monkeys that were trained with the BrainGate controlled the computer cursor on the first attempt.

The first three months of 2004 went by in a blur as we moved ahead
on many fronts: FDA approval, IRBs, publicity, recruitment, contracts,
the AAPM&R abstract, and other tasks. Cyberkinetics also asked me
to review a video about the BrainGate, which would be used to recruit
subjects.

At our research meeting on March 10, we learned that the FDA was
satisfied with our response. I circulated the approval letter from New
England IRB. Now all we needed to start recruiting patients was the
blessing of Rhode Island Hospital's IRB.

Gerhard said that the hospital was ready to approve the study, and
good news from the FDA and New England IRB would help resolve any
doubts. Late into the night, we discussed the protocol, public relations,
patient recruitment, and various regulatory requirements.

While driving home from Cyberkinetics headquarters, I reflected on
the information and felt overwhelmed. Burke had presented me with
a thick binder, including the final protocol, the investigator's brochure,
information for preoperative preparations, and adverse-event forms. It
was reassuring to have contact information for the research team as
well as Gerhard and his colleagues, but I wondered how a neurosurgical emergency would unfold. I patted the binder for reassurance, as one
would a companion before setting out on a long journey.

Shortly afterward, we received a favorable letter from the Rhode Island Hospital IRB. The study was officially ready for enrollment.

Publicity became a priority, and the research team was glad when *Forbes* ran a story in March 2004 titled "Mind over Matter," which proclaimed, "Once the realm of science fiction, neural prosthetics are slowly becoming reality." The reporter described John Donoghue's early "eavesdropping on the brains of rats and monkeys" with "primitive and unreliable" technology that required him to "glue a bunch of wires together and insert them into" the brain. Now it was time for the BrainGate. Tim was especially pleased with the favorable article, which lent credibility to his fund-raising efforts.

Everything had gone smoothly, but now we felt some pressure to recruit patients. In Rhode Island, we hoped to collaborate with PARI, an organization for people with disabilities. Leo Canuel, the executive director, was intrigued by the BrainGate, so I sent him some information and suggested a meeting with John Donoghue and Gerhard Friehs.

On another local front, the medical director of Rhode Island's long-term hospital for disabled persons was enthusiastic about the study. I e-mailed him the informed consent form and offered to give a presentation to the IRB and the state hospital's staff. That e-mail was sent on April 1, which was symbolic of my foolish attempt. In spite of sending information to the IRB, having a presentation almost arranged, and meeting all the requirements of the IRB, my efforts were futile. Perhaps wary of potential complications and more expensive patient care, the state hospital's IRB rejected the BrainGate study. Research can be an exercise in frustration, but we still hoped to recruit patients from other long-term care facilities.

Other problems arose. Shawn realized we didn't have a confidentiality form for compliance with the Health Insurance Portability and Accountability Act (HIPAA). When I faxed him a form from another study, he modified it and faxed it back. Then I faxed it to Sargent Rehabilitation Center, which approved a final draft.

For public relations, Burke commissioned a video shoot that included John Donoghue, Tim Surgenor, and me. To prepare, Burke and I discussed a list of questions I would answer to describe the BrainGate. Then the videographer took strategic shots of me staring at an MRI of the spinal cord, standing near an electric wheelchair, and walking in the gym. A voice-over was the final touch for this elaborate production.

Cyberkinetics managed to arrange an interview with Andrew Pollack of the *New York Times*. The reporter started off with a pointed question: Why did I want to be involved in the first human clinical trial? I explained that the preclinical data were convincing, that there were only a few minor complications, and that the BrainGate held great potential for enhancing the independence of persons with severe disabilities. Once a patient's neural signals could control a computer cursor, that could unlock a world of independence: e-mail, the Internet, household appliances, wheelchairs, and medical robots.

In the April 13, 2004, *Times* article, Pollack wrote that "the melding of man and machine has long been a staple of science fiction." Subjects "will have a cable sticking out of their heads to connect them to computers, making them look something like characters in 'The Matrix.'" Pollack cited Dr. Phillip Kennedy, who had implanted brain electrodes starting in 1998, enabling a severely disabled patient to type, although only at a rate of three words per minute. Jonathan Wolpaw, also a neurologist, was recording electrical activity from the brain's surface to control a cursor, as we were doing with the BrainGate.

Pollack noted that the BrainGate—which required an opening in the skin around the implant—carried a risk of infection, in contrast to Kennedy's brain implants that transmitted radio waves through skin that had been sealed by sutures. Miguel Nicolelis, a researcher at Duke University, was quoted in the article: "We don't like to hang around with wires coming out of our head." He also questioned the usefulness of the BrainGate, given our requirement that subjects should be able to talk; after all, they could already use voice-recognition systems or ones that tracked head or eye movements, such as the Eyegaze. His goal was to "control a robot arm, making three-dimensional movements," which he considered superior to the BrainGate's control of a computer cursor.

Also working on a brain implant, Richard Andersen at the California Institute of Technology said, "There is a feeling now that the time is here for moving the technology to testing humans." On the other hand, Dawn Taylor from Case Western Reserve University had some reservations: "a disaster at this early stage could set the whole field back."

Pollack carefully avoided overoptimism and noted that it was uncertain if "the implants will move around over time or cause scarring. Either could lead to loss of the neuron signal." He also asked if the BrainGate would be "superior to more mundane methods like voice recognition or even Dr. Wolpaw's" system.

Cyberkinetics responded that "its system would be faster than other methods, perhaps allowing people to type 20 to 30 words per minute, as fast as a healthy person could type on a BlackBerry." Mixing skepticism with hope, Pollack did mention that Cyberkinetics planned a miniaturized model that "would be fully implanted in the brain, transmitting information without wires. The price is likely to be in the tens of thousands of dollars."[5]

Kari Watson, our public-relations consultant, prepared a news release for April 20. During a media teleconference, Tim said, "This is an important neurotechnology milestone because the BrainGate system is the first example of a long-term human implant to provide a direct connection between neurons in the brain and a computer."

Later that afternoon, Felice Freyer from the *Providence Journal* called for an interview and arranged for a photograph to be taken of me near an electric wheelchair. Her front-page article—"Testing the Power of Thoughts"—was a far cry from the skepticism of the *Times*.[6] I was glad to see that "hope" was a prominent word. I tried to offer optimism and avoid pipe dreams. The story greatly helped our publicity in southern New England. Two local television interviewers featured the project without challenging my answers, perhaps because they were so taken with the technology.

Health care websites, notably CenterWatch.com and WebMDHealth .com, assisted us by posting a description of the BrainGate study. Media attitudes regarding the BrainGate study ranged from laudatory to provocative to confrontational. With its novelty and its potential for risk and reward, the BrainGate always drew a strong response.

<hr>

Soon after the interviews, I began to hear from people with severe disabilities. One of them was a person who had suffered a large stroke less than a year earlier; recent paralysis was an exclusionary criterion because further recovery was possible and could obviate the need for the BrainGate. Several family members also called, including the daughter of a person with multiple sclerosis, but this condition was not approved by the FDA for receiving the BrainGate. Some people qualified because of tetraplegia due to spinal cord injury, but they did not meet other criteria. For instance, I had to decline a person who was tetraplegic but older than sixty. Someone who had suffered a spinal cord injury twenty years earlier had adapted quite well and did not have significant computer experience, so it would not have been ethical to risk a decline in his status. Similarly, a person with muscular dystrophy was already fairly independent and able to type out e-mails. Inquiries from New Jersey and some Southern states did not meet the geographical criterion. I even received a request for information from a physician on behalf of a patient in Ireland.

After a flurry of e-mails and phone inquiries—but no suitable candidates—there was an unnerving lull as the media buzz faded. We sat and waited. I sent a letter approved by New England IRB to all neurologists in the area and received an informal referral, but this person did not meet two of the enrollment criteria. Another neurologist advised us to enroll patients with severe strokes and place the chip in the already damaged hemisphere. I wrote to Burke: "Meningitis or a large bleed could damage the other hemisphere. Since this is the first human study of a device with great potential, we should be extremely cautious to avoid any early disasters that could destroy our long-term goals." If patients with limited movement lost even that ability, I would feel terrible. My actions would be unethical, and I would risk a lawsuit as the principal investigator.

Michael Jones kindly sent a letter, which was approved by his IRB, to former Shepherd Center patients from New England. This letter did lead to a few calls, one from a young man who had recently suffered a diving injury. He had intact arm movement and some wrist extension. In spite of his impairments, he could perform most of his bathing, dressing, feeding, and grooming. He was functioning at a relatively high level, so I declined.

Burke wanted to carefully reevaluate our inclusion criteria, which appeared stringent in light of some inquiries. So our research team agreed to consider people with limited strength at the wrist or hand, in addition to those who could bend one or both elbows or lift one or both arms. Of course, this change in the protocol was made only after approval by the FDA, the Rhode Island Hospital IRB, and New England IRB.

At this discouraging stage, I was glad to see my interview published in the TechACCESS newsletter.

"What an exciting time we live in!" Carlson wrote. "Some technologies . . . seem to come out of nowhere and make you say 'WOW!' That is how I felt . . . about the amazing BrainGate neural interface system."

Play It Again

Kathy Spencer

“I wanted to pick up my grandkids. I wanted to hug my family with both my arms. And I wanted to play the piano again.”

Those were Kathy Spencer's priorities after a stroke paralyzed her right side at the age of fifty-three. Six years later, Kathy and I discussed her life since that day. A slim, petite woman with cheerful brown eyes, Kathy spoke excitedly and clearly while telling her story.

On October 23, 2005, Kathy had enjoyed a baby shower for Amy, her daughter-in-law, then dinner and a movie with her husband, Carl. When the nausea started, Kathy thought it was stomach flu. But soon her right leg cramped up, and her entire right side went limp. At the hospital, the ER doctor focused on her normal vital signs and her excellent medical history. He couldn't believe that she was having a stroke. But the paralysis did not improve, so he ordered a stat CT scan.

A stroke? An energetic woman who exercised four times each week and ate healthy food, Kathy was in disbelief.

“I'm going to start smoking and drinking,” she joked with her doctor. “It did me a lot of good to take care of my health.”

The doctor said that if she hadn't been in such good shape, the stroke would have killed her. An echocardiogram showed a plaque at her aorta, so she was placed on a blood thinner to prevent more clots from going to her brain.

Four days after her stroke, Kathy went for rehabilitation to Mount Carmel Hospital in Ohio. On arrival, Kathy had no movement of her arm or leg.

"I would have just fallen over," she told me. "Two people had to hold me up to stand."

Her right face drooped, and even though her mildly slurred speech was understandable, she asked for speech therapy.

"I couldn't say s and z clearly. I did a lot of public speaking for our chamber of commerce. So I told my speech therapist that I needed exercises for my face."

Because Kathy's right hand and wrist were limp, she was given a white plastic splint with soft Velcro straps.

"I slept in that for a year and a half," she told me. The device was meant to protect her from injuries and reduce the tone or spasticity that often develops after a stroke and can lead to a paralyzed, clawlike hand.

"The tightness started when I started weight bearing with physical therapy, about two weeks after the stroke," Kathy said. "When they gave it some exercises, my hand got tighter."

Kathy's hand went through the first two stages of recovery after a stroke—at first being limp and paralyzed, then having some tone. But her hand did not reach the third stage: regaining movement.

Her right leg came back much faster than her arm, and Kathy began walking with the help of a plastic brace molded to support her foot and ankle so that she wouldn't drag her lower limb and fall. It was the same kind of therapeutic process that Pat Wines had experienced, but Kathy's recovery was more rapid.

From the start of her medical journey, Kathy had all the support she needed. Carl stayed with her in the hospital, going home only for brief periods to take care of essential tasks and to feed their cat, Sunny. He also took notes and pictures during her therapies so that Kathy would continue her exercises with the proper form after going home. Her children visited regularly after work, until she said, "You don't have to be here every single day." But they all continued to come, offering encouragement and even giving their own therapy to help their mother's recovery. Amy would massage her right foot to ease its tightness and cramps. Her granddaughter, Jenna, who was in first grade, would move her arm back and forth ten times, then say, "Let's do five more." One day Jenna was sitting with Kathy on the bed, and she spontaneously kissed her right hand and foot and said, "I pray for your hand and foot every day." On Halloween, her family came in their costumes and brought a stuffed pumpkin.

Kathy's friends also played a vital role in her recovery. Fellow church members brought meals and offered encouragement. She got more than five hundred cards and so many flowers that she moved them into the common area for all the patients to enjoy. On occasion, her oldest son, Chad, had to serve as a gatekeeper to organize the flow of visitors. Another son, Adam, joked around with Kathy, getting her to laugh in spite of her condition.

Kathy also believed that her religious faith made a big difference in her rehabilitation.

"I'm a woman of faith, and it was crucial to my recovery. I got lots of prayers, and there was my daily praying."

In Psalm 137, the psalmist remembers Zion and imagines a stroke as a divine punishment for forgetting Jerusalem: "Let my right hand wither, let my tongue cleave to the roof of my mouth." In fact, some people told Kathy that the stroke was God's way of slowing her down.

But she replied, "God has put his loving arms around me and let me live because he had work for me. He wants me to help other stroke survivors." She believed that, with God on her side, her recovery would be a testament to divine intervention and to her praise of God.

During her challenging rehabilitation, Kathy often recited Philippians 4:13 to herself: "I can do all things through Christ, who strengthens me." Another favorite passage was Psalm 121, which offered a special resonance as "a song of ascents" and spoke directly to her disability:

> My help comes from the Lord,
>   the Maker of heaven and earth.
> He will not let your foot slip—
>   he who watches over you will not slumber;
>   indeed, he who watches over Israel
>   will neither slumber nor sleep.
> The Lord watches over you—
>   the Lord is your shade at your right hand;

the sun will not harm you by day,
nor the moon by night.

After twenty-four days, Kathy had recovered and adapted enough to go home. Her daughter, Karry, and granddaughter, Jenna, lived close to her and helped Kathy on a regular basis.

One of Kathy's good friends, Amy Baker, researched rehabilitation centers and recommended the Rehabilitation Institute of Chicago (RIC). Unable to get rehabilitation therapies five days per week in Ohio, Kathy went to RIC. The school where she worked as a counselor gave money to help with her expenses, and a friend, Jennifer Block, bought their airline tickets. At the Institute, she discovered a challenging and inspiring approach to rehabilitation.

"One of the first things they told me was to get rid of the cane, or my body would depend on it. I gave the cane to Carl and said I'd never use it again. At first it was scary because when anyone walked toward me I had to stop; I felt like I was going to fall. That was a very strange feeling."

I told Kathy that she had gone to one of the best rehabilitation centers in the world. I had interviewed there for my residency (but chose to go to Boston), taken the Institute's courses on braces and artificial limbs, and been involved in the rehabilitation of a patient who was treated there.

They stayed in Chicago for six weeks, though her insurance covered only twenty outpatient visits of physical, occupational, and speech therapy.

"The bills added up with the apartment and the therapies and everything," Kathy said. "But you can't put a price on your life."

Kathy told the therapists to give her everything they could, and if anyone canceled an appointment, she took the spot.

"They pushed me hard," she said. "There was another building where you could work out in between therapy visits. They gave me everything, even water therapy. I couldn't hold on to a floater or kick my feet very well, so the therapist held me."

Carl stayed by her side during the therapy sessions, to cheer his wife on and to photograph her progress. In one of her favorite moments, Kathy dangled in a harness to support her body weight, laughing while airborne. With half her body paralyzed, Kathy found that this offloading therapy was crucial for regaining her ability to walk without a cane and leg brace. (Some of my patients with strokes have benefited from a wheeled device called the Biodex; a therapist uses a winch to offload the weight and encourage a normal pattern of walking. It is similar to the wheeled walkers used by toddlers.)

At first, Kathy had almost no movement in her right leg, especially her foot. Her therapists at RIC started off with a simple exercise: they asked Kathy to scrunch up her toes to pick up a towel. It was three months before her toes started moving.

"I got a jar of seventy-five marbles," Kathy told me. "Every single day I picked them up with my toes and put them in a bucket, one at a time. I also did exercises with rubber Thera-Bands for my ankle. I'd lie on the bed, my husband would hold up my leg, and I'd try to keep it from flopping down. To strengthen my ankle, I wrote the ABCs in the air with my foot. When I was finally able to lift my leg myself, I put a one-and-a-half-pound weight around my ankle and lifted my leg up and down a hundred times a day." When her strength and balance improved, Kathy started walking on a treadmill with a metronome to help her gait become more normal.

"Rehab was a full-time job," she told me, "eight hours a day, seven days a week, week after week, month after month. It was well worth every minute."

Kathy used the plastic brace that supported her foot for only four months.

"I didn't want my ankle to depend on it. With the brace, I knew my ankle wasn't going to strengthen."

So she started walking around her house without the brace and also used the rubber Thera-Bands to improve her leg muscles. Soon she wore the brace only outside her house.

"Believe it or not," Kathy told me, "now my right leg is stronger than the left."

Early in her recovery, Kathy had prayed to God, "I just want to walk again. I don't care about my arm. I don't want to be in a wheelchair." Once she started walking, Kathy said, "Now, God, how about that arm? I'd like my arm back."

Kathy was right-handed, so using her left side was awkward. Even the skilled therapist at the Rehabilitation Institute of Chicago, who tried electrical stimulation, had failed to revive her hand. For half a year after the stroke, Kathy was unable to move her arm, wrist, and hand. So Carl moved it for her every day, neither of them giving up hope.

"What were you thinking when your hand didn't move?" I asked Kathy, wondering how she had managed to remain optimistic.

"I tried to stay positive. I was going to get it back. I told doctors not to tell me what I was not going to get back. I didn't want to hear that. I didn't want that in my mind. So I surrounded myself with positive things. My son, Matt, gave me motivational cards called 'Jesus' Keys for Success.' They had messages like 'Pray today,' 'Hope and cope,' and 'Believe and receive.' Even those cards were sending positive messages to my brain."

"How did you deal with negative messages?" I asked. I knew from my patients' experiences that they sometimes received well-intentioned but unhelpful advice.

"I've got to tell you this story," Kathy said. "In church, Carl and I always put our arms around each other's waist when we stand up for the hymns. One day a man asked why I didn't stand on the other side and use my good arm. I looked at him and said, 'I have two good arms. One's just weak right now.' I never sent any negative messages to my brain. I never called it my bad side."

"I tell patients it's the 'affected' side," I said. "And I tell them it'll get better—even if it seems unlikely. I'll never deprive a patient of hope."

Kathy laughed at my confession. "I'm glad you do. It's so important for the brain to hear that. Those messages are powerful. I tell people they're not stroke victims but stroke survivors."

"You have a great attitude, but did you feel discouraged at times?" I asked, wondering if her mind-set was too good to be true. I also wanted to know how she had managed to keep the stroke from muffling her musical laugh.

"Yes, sometimes. I was fortunate to not be clinically depressed. As a licensed counselor, I was familiar with that, but sometimes we can't see it in ourselves. So I asked my doctor at RIC, Dr. Elliot Roth, if he thought I was depressed. He laughed and said, 'If you were depressed, you wouldn't be working so hard and doing more than the therapists ask!'"

Kathy paused, as if remembering some difficult times, and said, "You have lots of losses after a stroke. You didn't just lose half your body; you've lost your independence, your job. . . . You have to mourn the losses and also work like crazy to recover, and it's scary because you don't know what you're going to get and what you're not going to get."

In spite of the paralysis that froze her right hand, Kathy believed that she had found the key to recovery: the piano, which she had started at the age of eight. Once, her fingers had played the black and white keys like sparrows fluttering over seeds; Kathy wanted her fingers to fly again. She began the healing process by listening to CDs of simple piano pieces every night.

"I'd lay my fingers flat on my leg. Then I visualized my right hand playing the melody of the same song over and over, day after day."

After seven months of these imaginary piano concerts, her index finger began to move, but just barely.

"A few weeks later, another one and another one moved. The last one was my ring finger." She still experienced tightness of the hand, however, so that prevented normal movement.

Nine months after her stroke, Kathy heard about a study at the University of Cincinnati to evaluate a functional electrical stimulation (FES) device called the NESS H200 Hand Rehabilitation System. This electronic brace helps patients regain the use of a hand affected by a stroke or spinal cord injury as well as other conditions. It supports the wrist in an optimal position for hand strength and coordination. Five electrodes are embedded in the custom-fitted brace to stimulate key muscles in the hand and forearm, resulting in hand movements that are useful for everyday tasks. Placement of these electrodes is individualized to ensure effective stimulation with minimal current.

The control unit is a miniature computer that allows the therapist and the patient to program stimulation patterns, depending on the type of weakness and tightness of the hand. The system comes with a battery charger, electrode pads that are periodically replaced, and a water applicator to help the conduction of electricity across the pads. The electronic brace is placed on the forearm and wrist, then connected by a cable to a control box, which the patient carries with a shoulder strap. A portable stimulator sends precisely coordinated electrical impulses to the forearm and hand muscles, restoring the ability to grasp and release objects.[1]

Because of this functional and repetitive movement, the computerized brace offers many other benefits. It reduces the tightness or tone that often occurs with spinal injuries and strokes, which can be severe enough to turn the hand into a useless and painful claw that has muscle spasms. Coordinated stimulation through the electrodes reeducates muscles to move in a more natural and efficient pattern. Functional electrical stimulation also counteracts the muscle atrophy caused by weakness and limited use of the hand. And muscle movement helps blood circulation in the hand, which results in less pain and a lower risk of skin breakdown due to pressure.

Clinical studies with the brace have shown improvements with activities such as bathing, dressing, and toileting. Therapy with the device can be started soon after a stroke or spinal cord injury and continued in the outpatient clinic as well as the home.

Much of the early research with the NESS H200 was conducted by Gad Alon at the University of Maryland. In an article published in 2003, he and his colleagues from rehabilitation programs in Sweden, the Netherlands, and Israel showed that patients benefited from using the device two or three times daily in a home-based, five-week program. Each of the volunteers had experienced a stroke more than six months earlier, had already undergone intensive rehabilitation, and had some movement of the wrist and hand. To assess the FES device, the researchers used tests that quantify the ability to grasp, manipulate, and release objects. For instance, in the timed Nine-Hole Peg Test, the patient places nine pegs into a board with nine holes, then removes them as fast as possible. Alon and his colleagues found that volunteers improved in all the tests, along with lifting objects and feeding. Most important for the quality of life after a stroke, patients had less pain and spasticity.[2]

Kathy had done enough research on the electronic brace to realize its potential for improving her hand movement, so she enrolled in the monthlong study. Before she started the treatments, a functional MRI assessed her brain's activity in relation to hand movement. The purpose of the pretreatment test was to see if her brain's firing patterns would change with repetitive electrical stimulation.

Kathy and Carl drove to Cincinnati and back five days per week. Her forearm and hand were fitted into the electric splint. The electrodes were moistened to improve the conduction of electrical current, and the therapist selected a program of patterned stimulation. For instance, during a sixty-minute session, one program might open the hand for thirty minutes, and another one might open and close the fingers and hand. Kathy described the sensation of stimulation as a little tingle.

"The first time it opened up my hand, I cried," Kathy told me. "Even though it's not under your own control, when you see any movement after paralysis, it's pretty big."

Throughout the treatment, Kathy felt a vibrating motion that relaxed her hand.

"It's very user friendly," she said. "You put the device on your arm, turn it on, and away you go."

Kathy believed that the device, through its repetitions of movements, retrained her muscles and her brain. At the end of the monthlong study, an MRI scan did not show significant changes in her brain, but her occupational therapist at Ohio State University saw improved strength and control of her fingers.

"Your movements are smoother, more natural," the therapist said. "You can pick up small objects more easily."

Convinced that the device had helped her, Kathy rented it and then bought one so that she could exercise her hand seven days a week.

"It's crucial to actively use it every day," she told me while we discussed the recovery of her hand. "It was the best money I've ever spent!"

"Yes, but this was many months after your stroke," I said. "How did your family handle this phase of your rehab?"

"You need to surround yourself with people cheering you on when you're going through something like a stroke," Kathy said. "My family and friends were very positive. After I started the NESS H200, Jenna would come over after school." The six-year-old girl had seen her grandmother's hand come back to life and regularly checked her strength. "You did better this time," she would tell Kathy. "Good job!"

Based on her experiences, discussions with therapists and physicians, and Internet research, Kathy developed her own theory of recovery after a stroke. Then she put it into practice. She performed countless repetitions of hand movements—associated with auditory, visual, and tactile sensory inputs—to stimulate her brain and to regain her hand movement.

"I'd use my hand and say what I was doing as I watched, so I was seeing, hearing, and doing the task. And visualization also helped my healing." While staring at the electric limb on her right hand, she stood in her pantry (see figure 11.1) and said out loud, "I'm opening my hand; I'm picking up the can of beans; I'm setting it down."

Kathy had intuitively discovered a basic principle of stroke recovery, first described by Thomas Twitchell in 1951: watching an affected limb during therapy helps improve the quality of movement.[3] If patients could not see their struggling hands, their movements were slow and awkward. Kathy was also in tune with more recent research, which showed the benefit of auditory feedback when patients are retrained to reach for objects while keeping their balance. She believed that her brain needed repetition in the rewiring of nerve pathways, but because her hand was still weak, the device would perform the necessary repetitions to jumpstart her recovery. A year had passed since her stroke, the time at which conventional medical wisdom advises patients that neurologic recovery will stop—but Kathy still planned to play the piano.

Four months later, with intensive self-rehabilitation and the electronic device, Kathy got full wrist action and was able to raise her hand against gravity, with the palm facing either up or down. Now it was time for strengthening exercises, so she used a one-and-a-half-pound weight and did a hundred wrist raises and curls twice a day. Before the stroke, she could have done ten times that weight. After seven months of using the brace for two hours per day, seven days per week, she opened up her hand and picked up a soup can—without the electronic brace. The intensive electrical stimulation had helped Kathy overcome the tightness of her hand, even without the device. She also picked up a juice glass, took a drink, and put it down.

"I advise people to jot down every little change in a journal," she said as we discussed her long course of rehabilitation. "The changes mean your brain is rerouting itself and you're healing and getting better. And celebrate every success along the way. That motivates you to keep going."

Ten months after starting therapy with the electronic brace and a year and a half after the stroke, Kathy had enough coordination and control to press a piano key with the isolated movement of one finger. Now confident about playing the piano, she tried a musically simple song that held great personal meaning. She played "Jesus Loves Me," concentrating on each finger. Then she called her friends and family to give recitals over the phone.

"It was great to play the piano, even though I was starting all over again."

When Kathy pushed down a piano key, her fingers tightened up, so she got a set of Oval-8 splints to keep her fingers straight. Even so, after a line of music, Kathy had to stop and massage her cramped hand. Gradually she built up her strength and endurance and began playing the piano for thirty minutes at a stretch. Kathy also moved nuts and bolts, one by one, from one wicker basket to another, working on her coordination, precision, and speed, which helped with the piano. Eventually, she advanced to playing one of her favorite songs, "O Holy Night," a hymn that provides spiritual and personal peace. In a video posted on YouTube, Kathy stands before her piano while telling her story, her right hand slightly more curled up than the left, and then she sits down and starts playing, her fingers graceful and melodic.

Kathy's leg had recovered much faster than her hand, but there were signs of recovery (in contrast to Anna Iacono, although I was still hopeful about her hand). Pat Wines and Linda Holmes regained enough hand movement for a variety of activities, such as horse riding for Pat and typing for Linda. In Kathy's case, the brace gave crucial electrical stimulation that initiated hand movement. It was as if the electronic brace had jump-started the process of hand recovery. In theory, Matt's hand could be made to move with surface electrodes in a brace or implanted ones in his muscles, but probably not on his own volition—unless the BrainGate could somehow be linked to functional electrical stimulation.

Kathy compared her recovery to the neurologic development of infants. She has eleven grandchildren, three of whom were born the year of the stroke.

"I watched them develop with a lot more intensity than the other kids. I felt like a little child. It was a whole new world, and everything was difficult. It was so bizarre. Everything is so different when you can't use your arm. It's something you can't even comprehend until it happens to you."

Kathy watched the children as they learned to roll in their cribs, hold up their heads, sit up, stand, and eventually toddle. Playing with them was a joy and an education. She saw a pattern similar to her own rehabilitation, as they first moved their entire hands and eventually achieved precise finger control.

"My grandkids took two years to develop," she said, "and I felt it would be the same for me. You have to give it at least two years."

"How're you doing at present?"

"My friends think I'm fully recovered because they can't tell I ever had a stroke," Kathy said. "I am so thankful! Living life is my therapy now. I can play the piano for as long as I want, with no tightness. Do I play as well as I used to? Not yet, but with practice, I'll get better and better."

To further improve her finger dexterity, Kathy writes every day and types on the computer for her work as a counselor. When she first placed the pen in her hand and tried to write, her words were barely legible. Now, six years after the stroke, her handwriting is almost as good as it used to be.

"I'm still fine-tuning myself and making progress," Kathy said. She uses intermediate-level Hammond books for piano exercises but hopes to fully regain her skills.

"With the brain's plasticity," Kathy told me, "we can improve until the day we die. We're never done unless we quit."

Plasticity and rewiring of nerve pathways! It was the first time I had heard a patient discuss the theoretical basis of stroke rehabilitation.

Kathy's determination in actually practicing this theory was even more impressive. I was reminded of groundbreaking research in the 1980s by Paul Bach-y-Rita, a pioneer in rehabilitation medicine. His chapter in a text on stroke rehabilitation,[4] which I read in the first year of my residency at Boston University, was inspiring. He defined brain plasticity as "the capacity to modify structural organization and functioning." Bach-y-Rita acknowledged his debt to Albrecht Bethe, the physiologist, neurobiologist, and physician who described the concept of neural plasticity in the early twentieth century. Bethe studied the microscopic anatomy of neurons and their connections; the electrical activity of nerves in isolation and in clusters; and regeneration in the central nervous system. In one of his experiments, Bethe examined shore crabs' brains after the amputation of one, two, or three limbs. Surprisingly, the crabs adapted to their disability and remained mobile, supporting the theory of neural plasticity due to brain reorganization.[5]

Early evidence for neural plasticity in human beings came from orthopedic surgery for patients with polio. To assist weak quadriceps muscles in these patients, surgeons relocated a muscle—the biceps femoris muscle (hamstring), which bends the knee—from behind the knee to the front. At first, this muscle continued to act only when the knee was bending, but soon the brain realized that the muscle should straighten the knee. Eventually the hamstring acted only to straighten the knee, like the quadriceps that it was meant to replace. These surgeons theorized in 1941 that the brain had adapted to the relocation of the muscle and realized that its new purpose was to straighten the knee.[6] This research supports the view that the brain can form new connections and create adaptive functions for the body.

Bach-y-Rita described another mechanism for recovery after a stroke: the unmasking of nerve pathways. As the brain develops, primitive reflexes are brought under the control of the brain's motor cortex, the highest level of neural function. For instance, an infant's stepping reflex when upright on a flat surface is replaced by attempts to walk when he or she reaches nine months. After a stroke, the nerve pathways for this reflex are unmasked, and intensive therapy to control the primitive movement patterns may help the patient walk.

At first during rehabilitation, a patient needs independence with his or her daily activities, so the patient is taught to use the unaffected arm. For instance, a shirt sleeve is first placed on the affected arm, and then the patient pulls it behind his or her back in order to put in the unaffected arm. This process involves a great deal of planning and wriggling as the patient struggles with a task that most people take for granted. Compensation for arm weakness is important, but the long-term goal should always be the recovery of function in the affected arm. That is the theory behind constraint-induced movement therapy, in which the unaffected arm is placed in a sling, and the weak one is forced to recover. Patients are carefully selected for this treatment, as there has to be some movement and sensory function on the weak side for the treatment to be effective.

Kathy told me about her intensive research into neural plasticity and its practical implications.

"It's a lot of hard work, but how much time and money is your life worth? God made amazing brains that can reroute. The Bioness lets us do repetitions to help that process, so we have to persevere with patience!"

Six years after her stroke, Kathy no longer used the device. Instead of therapy with the electrical splint, she played the piano, wrote, and typed. Kathy combed her hair and applied her makeup with her left hand.

"I tried with my right, but it's not nearly as good. I do use my right hand for the blow-dryer. After two and a half years of therapy, day in and day out, once you get some recovery you want to live again."

She continued to receive hand therapy on a regular basis but was working hard on other projects, so using her right hand for makeup was not a priority. Kathy found that everyday living was therapeutic: family activities, typing, writing, and working. She had two master's degrees, in school counseling and in school administration (with elementary principal certification).

"My work as a counselor helped my recovery," she said. "I knew about self-talk to motivate myself, about staying positive."

After the stroke, she studied clinical counseling at the Methodist Theological School and passed the National Counselor Examination in 2010. In 2011 she became a full-time counselor at Middle School West in the Gahanna-Jefferson Public Schools.

Kathy also began volunteering as a speaker for stroke support groups, and she gave me a sample of her pep talk:

> We need to help each other. We have to get our lives back. Don't let people do things for you. My friends hated it when I told them not to open the door for me; they wanted to help so badly. At first, I had to let my husband put on my gloves or tie my shoes when we went outside, but I did as much as possible. I thank God every day for my recovery. It's amazing what our bodies and brains can do. They used to tell people that you're done after twelve months, but you're never done. You're never done until you quit. First I relearned the piano, and now I'm teaching my granddaughter how to play [see figure 11.2].

The Antiseptic Baptism and the Neurosurgeon's Gift

Matt, May to June 2004

A while after the TechACCESS newsletter was circulated, the Greater Boston Chapter of the National Spinal Cord Injury Association referred Matt to us. His mother, Ellen, and I had a long discussion about the BrainGate study. Because of the tracheostomy tube in his throat, which resulted in hoarse and weak speech, a long phone call with Matt would not have been practical. So I mailed the informed consent form (ICF) to the Nagles. Now that the research team knew that this family was considering the study, we were optimistic but also nervous. Ellen's description led me to feel he met the study criteria, but would he agree to undergo neurosurgery?

With a prospective candidate outside Rhode Island, I needed a coinvestigator in Massachusetts to avoid the gray and risky zone of being a research doctor in a state where I had no license, especially when the complex licensure process could take a long time. As a faculty member at Boston University and Tufts University, I offered a research elective at my rehabilitation center in Rhode Island to residents. Gary Polykoff, a former resident who had coauthored a publication with me, was interested in serving as a coinvestigator. So I faxed New England IRB a new

FDA financial disclosure form, Gary's curriculum vitae, and his license information, along with a request to expedite the process.

New England IRB responded that he needed a Rhode Island medical license. This twist was entirely unexpected. A Massachusetts physician had to be a coinvestigator for a patient in that state, but Rhode Island licensure was also required because the study was officially based in Rhode Island—even though the physician would never see the patient in that state. Burke laughed at these convolutions; then he convinced the IRB that Gary required only a Massachusetts license.

A week had passed since I spoke to Ellen Nagle. She said that the family was still thinking about the study. I offered to speak directly to Matt, but she declined. At this point, my hopes were sinking lower and lower. I tried to be patient. Besides, I had no choice.

There were numerous other issues to address. My laundry-list e-mail to Burke on May 20, 2004, included the following: infection control, presentations to the PARI board and the state hospital system, mailings to neurologists and neurosurgeons, a television interview, Gary Polykoff, case report forms, and plans to highlight the role of Sargent Rehabilitation Center. As the official site for the first human study of the BrainGate neurosurgical implant, the center had taken a risk on behalf of people with severe disabilities and deserved a great deal of credit. Burke thanked me for the "tremendous support," and I felt glad to help the cause.

New England IRB sent me an e-mail with the final version of the ICF in mid-May. Around this busy and uncertain time, we were also happy to learn that the American Academy of Physical Medicine and Rehabilitation had accepted our abstract. Everything was falling into place—except for Matt Nagle's enrollment.

One evening after dinner, I received a page for a Massachusetts number. I called back, wondering if it was one of my colleagues from Boston University or Tufts or just a wrong number.

"Hello. Did someone page a physician?"

"Hello, Dr. Mukand," I heard a woman's voice say. Before I could exchange any pleasantries, Ellen said, "Matt wants to be in the study."

I couldn't respond at once because of my excitement about enrolling our first patient. After a deep breath, I collected my thoughts.

"Did Matt have any questions?"

"No," Ellen said in a matter-of-fact tone. "Everything seemed pretty clear in the form."

We discussed logistics, such as how to coordinate their vacation plans in July with Gerhard's surgical schedule. Preoperative testing included a CT scan, EKG, chest X-ray, and some lab tests. But the first step was to get Matt's official consent. Too many tasks were swirling around in my head, so I promised to call Ellen back with more details.

Thrilled about enrolling the first BrainGate patient, I left the house and wandered aimlessly around my Providence neighborhood, trying to calm down. Burke was just as excited when I called him. We basked in each other's compliments, chatted about how we had reached this point, and even joked about how the skeptics would react. Then we came back to earth and started planning the surgery.

I hardly slept that night, thinking of all the requirements of the protocol and the implications of this study for disabled people. Patients might actually use a neural prosthesis to control a computer by merely thinking about it. Of course, there was no guarantee that the BrainGate would work, but that thought only flitted across my mind.

Next morning, Gerhard and I discussed doing the surgery before his upcoming trip to Europe and mine to the Midwest, but it seemed prudent to wait.

In order to enroll Matt, I arranged privileges for Gary Polykoff at New England Sinai. Shortly after I faxed his curriculum vitae, license, and malpractice certificate, Gary was officially on staff. Now I could evaluate Matt and make clinical recommendations as the principal investigator of the BrainGate study and as Gary's colleague.

First, I had to arrange the screening visit with Matt. There was no guarantee he would meet all the criteria, even though he seemed like a good candidate. To expedite matters, I called Ellen at the school where she was a reading specialist. All the screening requirements demanded a lot of coordination, which became clear when we discussed possible dates when Matt could travel to Sargent Rehabilitation Center in a van designed for his electric wheelchair and portable ventilator.

My first task was to get Matt's clinical history. With Matt's permission via his mother, I called his physician, Dr. Brian Bloom. Intrigued, he wanted to help Matt enter the study, so he provided his medical history. I was surprised to learn of a recent possible seizure, but an EEG and a CT scan of his brain had suggested that the most likely cause was drowsiness from a medication and not a seizure. Gerhard felt that this isolated episode should not prevent Matt's enrollment.

The next issue was Matt's spasticity, but the involuntary movements of his limbs were mild and infrequent. As a result, Gerhard, Burke, and I decided that this should not exclude him from the study. On delving further into Matt's medical status, I found another hurdle: pain medications. Matt's shoulder and neck pain required narcotics that could make him drowsy and interfere with the BrainGate study.

Dr. Bloom discussed this obstacle with Matt, and they agreed to reduce his sedating medications as much as was reasonable. Later, I learned that Matt had reviewed the study criteria and started reducing his pain and spasticity medications even before our screening appointment. Gerhard, Burke, and I felt that these drugs were not major issues and would not be considered deviations from the protocol. Matt was delighted to learn that he had sailed through the first stage of his enrollment, in spite of turbulent waters.

The second stage was more challenging. Matt had a history of methicillin-resistant staphylococcal aureus (MRSA) infection. This drug-resistant organism is common in long-term care facilities. People can be colonized for years without an active infection, but if the bacterium attacked Matt's brain via the surgical site, that would be life threatening.

Worried about this potential pitfall, I called Dr. Leonard Mermel, an infectious-disease specialist at Rhode Island Hospital and a consultant for Cyberkinetics. He suggested cultures from various body areas. If necessary, we should treat Matt with antibiotics and total body baths with the antiseptic Hibiclens, followed by another set of cultures. While planning the study, the research team had agreed that MRSA would exclude a patient from the study.

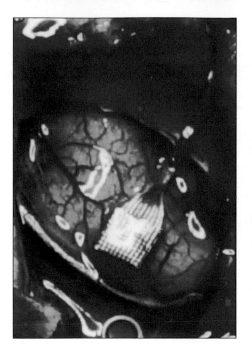

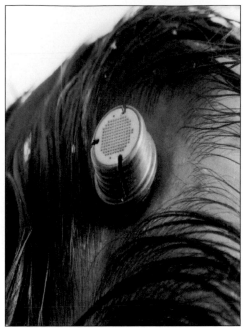

(top two photos) 1.1 | The BrainGate implant in Matt's brain.
Courtesy of Cyberkinetics

(below) 2.2 | Matt playing hockey.
Courtesy of Ellen and Pat Nagle

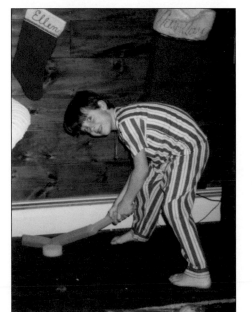

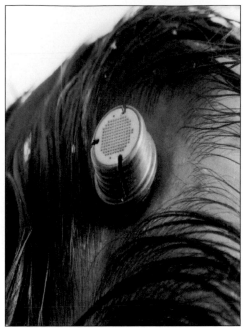

(above) 2.1 | A horizontal cross section of the spinal cord shows the descending motor nerves, the ascending sensory input from peripheral nerves, and the butterfly-shaped gray matter.
Courtesy of www.thebrain.mcgill.ca

(clockwise)
3.1 | A homunculus represents how specific areas of the brain control the movement of various parts of the body. *Courtesy of www.thebrain.mcgill.ca*

3.2 | Anna Iacono at the staircase. *Author's collection*

4.1 | Matt in football uniform. *Courtesy of Ellen and Pat Nagle*

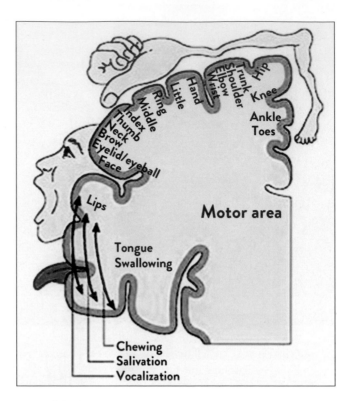

5.1 | Matt with his friends and Britney Spears. *Courtesy of Ellen and Pat Nagle*

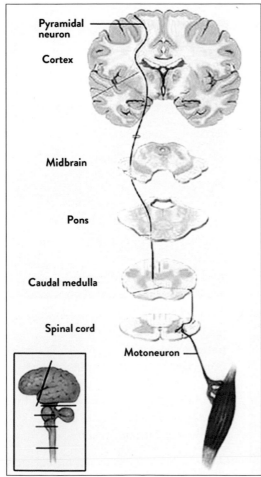

(*left*) 6.1 | Cross sections of the central nervous system start with the motor cortex at the highest level (inset). The upper motor (pyramidal) neuron descends from the cortex to the spinal cord and controls the lower motor neuron (labeled "motoneuron") to stimulate the muscle fibers and cause movement. *Courtesy of www.thebrain.mcgill.ca*

(*above*) 8.1 | The microelectrode array that recorded the electrical activity of individual neurons. *Courtesy of Cyberkinetics*

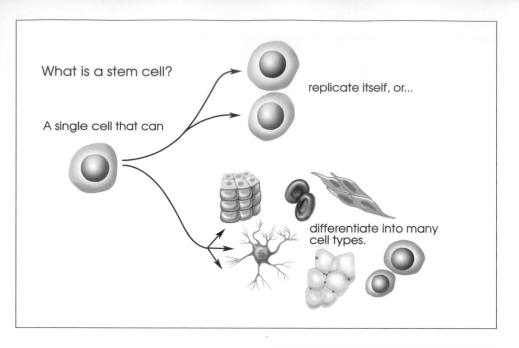

What is a stem cell?

A single cell that can

replicate itself, or...

differentiate into many cell types.

(above) 5.2 | Pluripotent stem cells differentiate into specialized cells or replicate themselves.
Courtesy of the National Academies Press, Washington, DC

(above) 6.2 | The L300 on a patient's leg.
Copyright © 2011 Bioness Inc.

(left) 6.3 | Pat with her granddaughter, Madeline, and her horse, Comet.
Courtesy of Patricia Wines

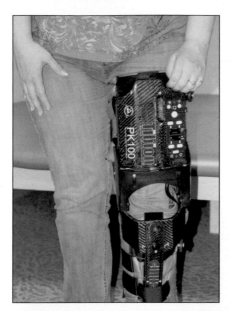

(top left) 9.1 | Linda Holmes wearing the bionic leg made by Tibion.
Courtesy of Whittier Rehabilitation Hospital

(top right) 11.1 | Kathy Spencer practicing with the H200.
Courtesy of Kathy Spencer

(left) 11.2 | Kathy Spencer and her granddaughter Jenna.
Courtesy of Kathy Spencer

12.1 | Matt's MRI before the implant surgery, showing a horizontal cross section.
Courtesy of Ellen and Pat Nagle

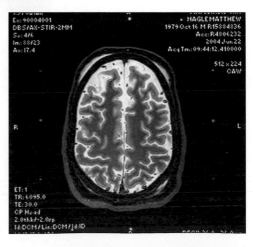

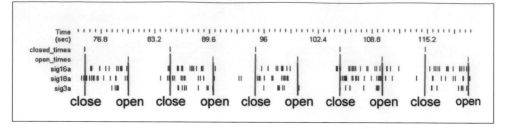

Time (sec)	76.8	83.2	89.6	96	102.4	108.8	115.2

closed_times
open_times
sig16a
sig18a
sig3a

close **open** **close** **open** **close** **open** **close** **open** **close** **open**

(above) 13.1 | Neuronal activity increased in three neurons when Matt imagined closing his hand.
Courtesy of Cyberkinetics

(right) 13.2 | Connected to the BrainGate computer by a pre-amplifier box and fiber-optic cable, Matt moved the cursor by using only his thoughts.
Courtesy of Cyberkinetics

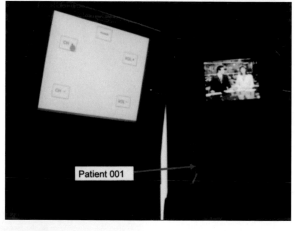

(right) 13.3 | The BrainGate read Matt's brain waves, and he was then able to select the box on the screen to turn the television on or off, adjust the volume, and change channels.
Courtesy of Cyberkinetics

Patient 001

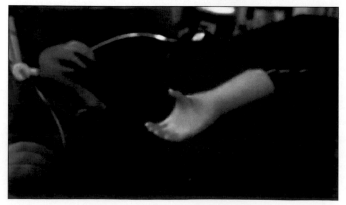

13.4 | With the BrainGate interpreting his nerve signals, Matt opened and closed a prosthetic hand. This task was a precursor to controlling a robotic limb.
Courtesy of Cyberkinetics

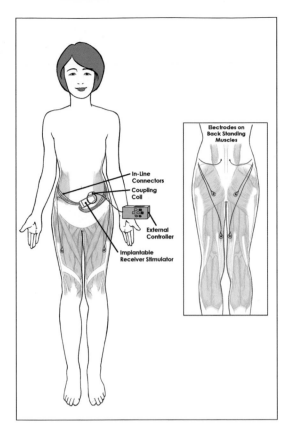

(left) 13.5 | A functional electrical stimulation (FES) system similar to that used by Jennifer French.
Courtesy of Cleveland FES Center

(below) 14.1 | The mPower1000 is the latest version of the Myomo and is FDA approved for home use.
Courtesy of Myomo

13.6 | Jennifer French sailing.
Courtesy of Jennifer French

14.2 | Garrett on the ice.
Courtesy of Eileen Mendez

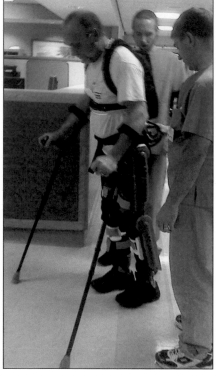

17.1 | Floyd Morrow using the ReWalk.
Courtesy of MossRehab

19.1 | Matt with
his cousin and his
cousin's son.
*Courtesy of Ellen
and Pat Nagle*

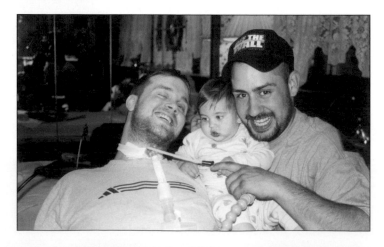

As feared, the cultures showed that Matt was colonized with MRSA. An infection could travel along the implant wires to the brain membranes, resulting in meningitis, or to the brain itself, causing encephalitis. I shuddered. That could cause seizures, coma, or death. MRSA kills about twenty thousand people in the United States each year.

Matt received oral antibiotics, five consecutive daily baths of Hibiclens, and Bactroban ointment at the affected sites. We purified Matt by baptism in antibiotics, but the cultures still showed MRSA at one site. It would probably never be eradicated.

Panicked, I called our consultant, this time with Gerhard and Burke on the phone. I asked if we could use antibiotics such as intravenous vancomycin and topical Bactroban at the time of the surgery.

"You can do that, but we shouldn't use him as the first patient in such a high-visibility trial." Dr. Mermel was understandably cautious. "Let's treat him again."

The second round of antibiotics also failed. Now we had only two options. One was to bathe Matt with Hibiclens for three days before surgery, give intravenous antibiotics, and paint his scalp with brown Betadine antiseptic. Meticulous postoperative wound care would be essential. Our other option was to cancel the BrainGate implant for Matt.

As the principal investigator, I discussed the choices with his mother and arranged to speak directly to Matt. He had no concerns about the risk of infection.

"I'm all set, Doc," he said hoarsely, while the nurse held the phone. Ellen went along with Matt, as she usually did.

On May 25, the day of Matt's appointment, I arrived early at Sargent Rehabilitation Center. On the way to my office, I walked past an antique wooden wheelchair on display, its polished oak frame latticed with woven wicker to distribute pressure and prevent potentially fatal pressure ulcers. A functional work of art, this chair was more comforting and aesthetic than modern metal wheelchairs.

While I prepared the informed consent forms, a nurse from New England Sinai called to say that a nurse's aide would accompany Matt—instead of his parents.

As the implications dawned on me, I was silent. Matt was coming for a crucial screening appointment to become the first human BrainGate subject. He had probably convinced his nurse's aide that he would enroll in the study—regardless of what I said. It was essential for his parents to serve as advisors, to temper his enthusiasm with awareness of the risks. Because the study involved travel, his family would be responsible for transportation and other logistic details. Without their support, the study would be doomed.

"He should stick with the plan and come with his parents," I said to his nurse.

Was there some friction within the family regarding the study? I had recently confirmed all the details with his mother, but this sudden change of plans was not a good sign. Perhaps Matt was trying to control one of the few areas of his life that he could influence.

Nonetheless, I didn't want to take advantage of a severely disabled young man by enrolling him without the presence of his parents. For another study, I would have gone along with his wishes—but not the first human study of a brain implant.

His nurse went back and forth between us and explained why his family's involvement was essential. At this point, I was fully prepared to keep Matt out of the study. This situation did not bode well for future collaboration with the subject and his family.

I waited for Matt's response via his nurse. On the threshold of the most risky clinical venture of my career, I felt my fears about the Brain-Gate study bubble to the surface. As the principal investigator, I would be linked with adverse outcomes, such as an infection, seizures, or death. I imagined comments about taking excessive risks for the sake of research. I imagined people questioning my motives. I imagined the consequences for my career. None of this helped my anxiety. The media had already heralded the BrainGate study, and a clinical disaster would be a career disaster. So I was glad when the phone rang and the nurse said that Matt would travel with his parents. I thanked her and finished preparing the forms.

||||||||||||||||||||||||||||||||||||||||

I waited by the elevator for Matt to arrive. Even after twenty years as a doctor of rehabilitation medicine, I am careful about interacting with people with severe disabilities. The first social barrier is the difference in height between someone in a wheelchair and a standing person. Although a wheelchair gives mobility and independence, it is also a tangible sign of disability: a black vinyl and chrome machine attached to a human being, a tragic symbol that lurks below the surface of social discourse. Then there is the initial awkwardness of greeting a person with tetraplegia; a hand gesture is definitely out, because the patient cannot wave back. A similar clumsy moment occurs because it is impossible to shake hands. My approach is to take the patient's right hand with both of mine, hold it briefly, and sit down as quickly as possible to face the person in the wheelchair and avoid any further social discomfort.

Matt and I had never met before that day. We had communicated only by phone and through his mother and nurses. He had been a star football player, so I expected a young man who was well built. Beyond that, I had no idea what Matt would look like.

When the elevator opened, the young man in the motorized wheelchair appeared. He was wearing a New England Patriots cap, either because of team spirit or because he hadn't had his hair washed recently. When I introduced myself, Matt grinned and winked, leaving open a glinting blue eye. A normally reserved person, I was disarmed and grinned back—but didn't wink.

Later I realized that Matt was committed to the BrainGate study and had already established a mental relationship with me as his "research doctor." I had expected a muscular, square-jawed young man, but Matt's physical condition had declined, as was visible in his rounded face and his lack of abdominal and chest musculature. Paralysis had softened a physique sculpted by weight lifting, football, and hockey.

Matt was seated on his large black Arrow wheelchair, a three-hundred-pound carapace with a battery and portable ventilator that allowed him to be mobile. I held his hand in mine as a greeting, and he forced out a hoarse "hello."

While Matt inhaled and exhaled through a sip-and-puff tube to maneuver his wheelchair, I made small talk with his parents to suppress my nervousness about this uncharted journey. Pat Nagle was a tall, dark-haired man with a strong physique. Ellen was a cheerful woman with short blond hair and a warm smile. Inside the office, Matt positioned his wheelchair in a corner, his parents sat at the other end of the table, and I was in the middle. Safely parked, the chair beeped three times to signal it was in neutral.

Matt appeared ready to give his full consent, but I was ethically and legally required to discuss the study, procedural details, and possible complications. So, though the informed consent form was fourteen pages single spaced, I read every word to him. After each section, I asked Matt if he understood the materials or had any questions. It became obvious that he had carefully read the entire form, and he frequently nodded his head in assent. Matt's parents, especially his mother, would periodically join in the discussion of what Matt had to hear, understand, and accept.

I explained that the BrainGate system was an investigational device that did not provide any treatment and would potentially allow control of a computer cursor by using only thoughts. An implant would monitor the electrical activity of brain nerve cells that controlled hand muscles. The implant was connected by a bundle of fine wires to a pedestal that protruded through the scalp, then to the computer by a fiber-optic cable. The BrainGate computer interpreted the brain signals, which might allow control of the computer cursor by only thinking about it.

The long list of inclusion criteria required understandable speech, intact cognition, good mental and physical health, no other implanted devices, weekly visits for one year, and computer experience. I repeated that screening tests included an EKG, a chest X-ray, a CT scan of the brain, and lab tests for the blood cells, electrolytes, and measures of blood clotting.

Then I described the surgery, which would take four to eight hours. Standard postsurgical care would include a CT scan of the brain and routine physical examinations and blood tests. Follow-up care would be provided by the nurses at New England Sinai. After Matt's skin healed, the BrainGate would start to record his nerve signals. Then Matt would be trained to use the BrainGate system to move the computer cursor by thinking about it. If he wished, Matt could take breaks. Testing would be limited to ten minutes at a time. He might be photographed or video-taped for the FDA.

I advised Matt that if there were problems with the implant, it might have to be removed. According to the protocol, the electrode array would be removed at the end of the study, after thirteen months, in a surgery similar to the implant procedure. Cyberkinetics would pay all study costs. Matt would not be paid for this research. He did not have to participate in the study, because other assistive technology was available.

Risks of surgery included infection, bleeding, pain, swelling of the brain, stroke, neck cramps, fever, skin problems, nausea, seizures, and death. Having the implant in the brain created additional risks: headache, loss of head and neck movement (for he had no other movement to lose), spasticity, swelling of the brain, confusion, speech or swallowing problems, visual changes, implant failure (assuming it worked), and death. There was no assurance as to how well the BrainGate system would work, if it worked at all, but I noted that, in the monkeys, the system had worked for an average of sixteen months and as long as thirty-three months.

If the results were promising, Cyberkinetics could extend the study. Matt had to allow me to get medical information from any and all sources, but his records would be confidential and protected.

Matt could always leave the study, and this would not affect his regular health care. Conversely, I could terminate the study if Cyberkinetics or I felt it was in his best interest, and the BrainGate implant would be removed. Cyberkinetics also had the option of ending the study at any time for reasons unrelated to Matt's health.

If Matt were injured due to the study, Cyberkinetics would pay for fair and proper treatment, but not for pain or suffering. In the event of his death for any reason, Cyberkinetics would request removal of the BrainGate implant, along with a small portion of his brain, for further study. He and his caregivers had to agree to follow the recommendations in the "Patient and Caregiver Manual."

For any issues with the study, he could contact New England IRB. Gerhard Friehs would perform the surgeries because he had done most of the monkey surgeries. I disclosed Gerhard's conflict of interest as a founder and stockholder of Cyberkinetics and mine as a paid principal investigator for the BrainGate study.

Half an hour later, Matt still did not have any questions, and I realized that his decision had been irrevocably made before he arrived here. I didn't have to deflate the cuff around the plastic tracheal tube to increase Matt's air flow and allow him to speak. On the one hand, I was pleased that there were no questions or doubts, but this kind of certainty could turn out to be a problem. I tried to interpret Matt's determination in a positive light. Like many people with severe disabilities, he was unable to make many decisions for himself regarding even basic activities that are usually taken for granted: brushing his teeth, eating breakfast, and urinating as needed. He deserved to have some control of the research process.

At this point, I realized that the BrainGate could be therapeutic for Matt, because the plastic tube that connected him to a ventilator prevented the use of a voice-activated computer system. Consequently, I felt much better about allowing him to take the risks inherent in the implant.

Matt's parents signed to certify that he had read the informed consent form, that all questions had been satisfactorily answered, that he fully understood the study, and that he agreed to participate *voluntarily* in this study.

Then Gerhard Friehs arrived with his consent form, which was shorter but included surgical details. A tall, husky man who looked like a nose tackle for the New England Patriots, Gerhard was wearing a suit but spoke informally to the Nagles. When he described how the skull bone would be sawed after peeling away the overlying skin, Matt's neck arched back slightly. I looked from Matt to Ellen, who stared out the window at the perfectly blue sky, then to Pat, who looked grim. Matt said nothing after Gerhard paused, and a gust of silence blew into the room. Matt swallowed, and the tracheostomy tube moved, as though helping him catch his breath.

Prepared to file away the seven copies of the consent form and call it a day, I waited for Matt to speak. Then the corner of his mouth curved up. Was it a resolute smile, was it bravado, or was it Matt's physical response to fear?

"You don't have to do this, buddy. We can go home," Pat said.

"Let's do it." Matt's chin quivered slightly. "I'm ready."

He looked over at me and nodded.

"Are you sure, Matt?" I said. "This is a big decision. You can think about it for a while."

"I'm sure." Matt confirmed his willingness to be the first BrainGate research subject with a wink.

With the paperwork finished, Gerhard and I learned that our subject was ambidextrous but left-handed with computer tasks. So the implant would be placed in the right brain.

While enthusiastic about the project, I was keenly aware that Matt didn't absolutely qualify for the study due to certain exclusionary criteria: sedative use, MRSA, and autonomic dysreflexia. I knew of the first two problems, but the third had only become apparent with my detailed and pointed questions. Matt's sympathetic nerves could become hyperactive and cause sweating, headaches, and dangerous surges in blood pressure.

Sensing my reservations, Burke suggested that I document our agreement on allowing him into the study. If I enrolled Matt on my own, in the event of an adverse outcome I would have to explain to his family, Sargent Rehabilitation Center, New England IRB, Rhode Island Hospital IRB, and the FDA why I had flouted the study criteria. Cyberkinetics had indemnified me to the tune of a few million dollars per patient, but even the remote prospect of a malpractice suit was disquieting.

In a long e-mail to Gerhard, Burke, and Shawn, I requested their agreement that "clinical judgment" could be used to loosen some of the following criteria.

First, chronic sedative use was a criterion for exclusion. Matt needed Valium for spasticity, but he was not sedated and his mental-status examination was almost perfect. In addition, because Matt was so intent on the implant, he had asked Dr. Bloom to taper off medications similar to Valium, such as Xanax and Ativan.

Second, only one episode of autonomic dysreflexia had occurred in the past three months, related to a bladder problem. This was a very low frequency, and the source of the dysreflexia was clear; therefore I felt it should be manageable in the future.

Third, we had tried to eradicate Matt's MRSA, but one area was still colonized. I left the decision regarding surgery up to Gerhard, as the surgeon should have the final say. With his approval, we would proceed and use preoperative intravenous and topical antibiotics as well as antiseptics.

Having reached an agreement with the key members of the research team, I scheduled the surgery for June 22, 2004, after Gerhard and I returned from our vacations.

In the interim, Matt and I talked several times. His room in New England Sinai had large windows with a sunlit view of a tall white pine on the lawn. Most of the space was taken up by a large electric bed with inflatable sections to protect his skin; his attacker's knife had cut all the sensory nerves in his spinal cord, so he did not feel pressure or pain, and his skin was always at risk of developing ulcers and infections. A large poster collage of friends' photos was titled "We Love You Forever, Matt Nagle." In a cardboard box were medical supplies: wound dressings, bladder catheters, and tubes to suction his lungs. A collection of movies included *Rocky*, which tells a classic underdog story of a boxer overcoming insurmountable odds. An electric shaver and toothbrush stood in their battery chargers, ready for the nurse. On his bulletin board were photographs of Matt with his family, scraps of paper with phone numbers, and menus from Slices Italian Pizzeria and Ninety-Nine to Go.

During one of my visits, Matt described the stabbing: "I was dead for an instant or an hour, I don't know how long. Katie held me as I turned blue and died. My lungs collapsed, but the paramedics saved me and got me to Boston Medical. Someone said, 'Come on, man, pull through,' and I'm glad I did. I tell my parents I love them every time I see them. You can't take anyone for granted, you can't take life for granted—life's too short. Too short. . . ." Matt stared down at his pale and paralyzed right arm, held down by padded Velcro straps.

"You're right, Matt." I placed my hand on his shoulder, where he could still feel the warmth of another human being.

Once the surgical date was set, I had numerous conversations with Matt's medical staff to get his current medical status and recent lab work and cultures. If there was the least complication, I was prepared to postpone the BrainGate surgery or even to withdraw Matt from the study. Most important, I tried to prepare Matt's parents for the surgery—and possible complications. Everything was proceeding on schedule, so I went off on my family vacation.

In spite of all our planning, some purely logistical problems arose. Two days before surgery, our ambulance company in Rhode Island said that it had no employees licensed in Massachusetts. Even a high-tech, cutting-edge research study has to contend with mundane details. Luckily, New England Sinai coordinated Matt's ride to Rhode Island Hospital. On the way to Providence, an emergency technician watched Matt's heartbeat racing across a green monitor. A sensor clipped to his thumb showed an oxygen saturation of 100 percent as the ventilator pushed in almost a liter of air with each breath, twelve times each minute.

Having finished my work at the rehabilitation center, I went to Rhode Island Hospital on June 21 and asked Matt for the last time if he had any questions.

"No." Resolute as ever, he smiled.

I smiled back. "How was the ride?"

"Fine," Matt whispered through the tracheal tube.

Some headphones were lying on his bed. To dispel the anxiety in the air, I asked, "Matt, what were you listening to in the ambulance?"

"Led Zeppelin's 'Stairway to Heaven.' Great song, great band."

"Great lyrics, too," I said. "My favorite part is when Robert Plant sings, 'A new day will dawn for those who stand long, and the forests will echo with laughter.'"

"'And it makes me wonder . . .'"

<hr />

The next morning, calmed with Ativan and connected to an IV pump and a portable ventilator, Matt traveled down an elevator, through a maze of hallways, and into a basement corridor with scuffed off-white linoleum and Impressionist paintings. Visible through the basement window were two tall chimneys rising into the dawn. At the large double doors to the open MRI suite was an orange warning sign: STRONG MAGNETIC FIELD— NO PACEMAKERS, METALLIC IMPLANTS OR NEURO-STIMULATORS. We had worried about the knife tip embedded inside Matt's spine—would the huge magnet pull it out or twist it deeper inside? A radiologist approved the scan because the Open MRI, with a less intense magnetic field than the standard MRI, was designed for neurosurgical patients with metal implants in their brains or spines. Its larger opening also allowed two long hoses to go from the ventilator to the white plastic tube in Matt's throat.

Matt was wheeled into a bright corridor, past a metal cabinet with X-ray files and a large laundry cart with a blue vinyl cover. On the wall to his left was a sign with instructions for treating bronchospasm, hypotension, tachycardia, and bradycardia. Matt didn't say anything and stared up at the white ceiling tiles.

The Cyberkinetics team was waiting for him. Gerhard wore a green scrub suit, ready for neurosurgery, and smiled reassuringly at Matt as he reviewed the procedure. John Donoghue, the father of the BrainGate and chairman of the neuroscience department at Brown University, beamed at Matt as though he were a star student. Burke greeted Matt politely and kept a professional distance as a representative of Cyberkinetics.

Gently, I touched Matt's shoulder.

"How're you feeling?" I asked, with my eyebrows raised ever so slightly. As we had discussed when Matt gave his informed consent for the study, he could back out at any time. This was his last chance before Gerhard fixed the stereotactic surgical frame on his head.

"Fine, Doc. Wake me up when the scan's over."

Matt's ventilator hushed his voice. Both Matt and Gerhard had no doubts; I wished for that kind of certainty.

At the Open MRI room, a combination lock prevented unauthorized entry, and a sign warned: DANGER: THIS MAGNET IS ALWAYS ON. Forbidden items included metallic prostheses and implants, firearms, scissors, keys, and watches.

A technician brought over the stereotactic Leksell frame, an oval ring with four posts; it weighed less than two pounds and was made of an aluminum alloy to minimize interference with the MRI image. Each side of the frame was marked in millimeters to precisely locate structures in the brain. Gerhard cleaned off four spots on Matt's head with Betadine swabs, anesthetized them with lidocaine, positioned the halo ring, and then tapped the screws into his skull. As he tightened the four posts, Gerhard asked Matt if he was OK.

Matt whispered, "Yes."

In the operating room (OR), a semicircular arc would be attached to the pieces projecting from both sides of the frame, to precisely locate the site for implanting the BrainGate sensor.

Connected to a portable ventilator that withstood magnetic forces, Matt was wheeled over to the Siemens MRI machine, a square white box the size of a small room. A large black hose rose from it into the ceiling to vent oxygen and helium gases. The box was attached to a platform with a wooden bed that was hollowed out to prevent movement. Four technicians transferred Matt onto the platform and slid him along the tracks so that only his head was in the doughnut's chamber. Then the heavy door to the MRI suite was shut. Matt was alone in the magnetic chamber.

John, Gerhard, the radiology technician, and I gathered in the small, dimly lit adjacent room with a window to watch Matt. On a shelf were supplies such as IV catheters, tape, and syringes. A gray box labeled TOX-GARD monitored the oxygen level in the MRI room, and near it was a contrast-reaction kit for allergic reactions to intravenous dye. A small music console was silent.

A switch on the computer keyboard allowed communication with Matt. His head was perfectly positioned, and we had allowed him to relax with the help of the Ativan.

"Can you adjust the contrast?" Gerhard asked the technician.

On the horizontal cross sections of Matt's brain, we were pleased to see a prominence in Matt's right brain, the outer gyrus where the Brain-Gate sensor would be implanted (see figure 12.1). Cautious by nature, I had suggested a functional MRI, which would light up the brain area when Matt imagined moving his hand. But Gerhard and John felt certain that the topography of Matt's brain would allow accurate location of the hand motor cortex based on neuroscience research from the past five decades.

Crowned with the Leksell frame, Matt was transferred onto the gurney and wheeled up to the OR.

I gave him a thumbs-up signal and said, "See you upstairs."

Matt nodded once. He was all set for the surgery.

In the waiting room, Ellen sat in a chair with salmon-colored vinyl upholstery near a table with some magazines. On the opposite walls were paintings of a harbor and a New England home. When I looked around for Pat, she said that he would arrive soon. I told Ellen about the MRI results.

"Thanks," Ellen Nagle said, without any facial expression, but her eyes sparked with fear. This was only the first step.

"Matt's on his way up to the OR, and I'll meet you after the surgery." I took Ellen's cold hand in mine and said, "I've worked with Dr. Friehs for years. He's a terrific neurosurgeon. And he's prepared as much as we can for this procedure. Did anyone tell you about the cow brain and the cadaver?"

Ellen drew back in her chair. *"What?"*

"Well, first Gerhard practiced the bone flap and the implant in a plastic model of the skull. Then Shawn went to a butcher in Boston and picked up a cow brain for Gerhard. And John Donoghue arranged for Gerhard to get more experience with a cadaver at the Brown Medical School. On the MRI, Gerhard saw exactly where to place the BrainGate." I tapped the approximate area on my own skull. "It gives a three-dimensional picture of Matt's brain, with x, y, and z coordinates so we can locate the exact place for the implant."

"But I worry, I worry. . . ." Ellen's voice became teary.

"Don't worry." I showed her a picture of the Leksell frame that had been fixed on Matt's skull.

Ellen's grip tightened on my hand. "I hope this works. He's been through so much."

"I know." I placed my arm on her back for moral support. "If it makes you feel any better, a lot of people are worrying about Matt today."

Many scientists, as well as people with severe disabilities who closely followed their research, were aware of the high-stakes marathon to develop the world's first brain-computer interface. Matt wished to enter this gateway to his brain, to allow his nerve signals to be recorded, interpreted, and used to control a computer. Having lost nearly all movement of his body, Matt yearned for any control over his existence, through any pathway.

Shortly after 8:00 AM, surgeons, nurses, and technicians were relaxing in the OR lounge. Some had coffee while discussing their cases; others read the newspaper. A lot of the surgical action was already going on or about to take place. From a cart I grabbed a pair of blue scrubs, shoe covers, and a cap and entered the locker room, where old surgical paper boots and masks were scattered around. Unused to surgical garb, I felt cold, and the hairs on my forearms stood up.

In the OR hallway, I found Burke and a technician from the engineering, research, and manufacturing facility of Cyberkinetics in Utah. Both were wearing their scrub suits, ready to bring in the BrainGate implant after Gerhard had sawed through Matt's skull. Burke opened the black, insulated case that held the steam-sterilized implant. Clinical engineers at Rhode Island Hospital had ensured that there were no electrical problems such as an improper ground or faulty wiring. The microelectrode sensor had a bundle of gold wires leading to the pedestal, which looked like a stainless-steel bolt, but Burke preferred that I not use that term.

"Jon, we're not working with cars, you know."

"Sorry, Burke. At least I didn't call the connecting piece a nut."

The vice president of clinical operations sighed at my irreverent comment, but the technician grinned.

Matt was lying on the narrow Skytron table in the operating room. Nearby were a blue garbage can for the disposal of paper gowns and wrappers of surgical equipment, a red can for biohazards, and a laundry basket. A computer was ready for the circulating nurse to record details of the surgery, such as the names and doses of medications and when they were administered. The OR nurses rotated the table so that Matt's right arm was easily accessible to the anesthesiologist for IV access. The Apollo anesthesia machine, named after the Greek god of healing, stood ready to deliver oxygen and anesthetic gases to Matt. Three computer screens would display Matt's status during the three-hour surgery—one on the main computer, a second on a side arm, and a third on top of the Apollo.

Laid out on a small tray were the anesthesiologist's instruments, including a plastic face mask and syringes. On a Pyxis cart for medications, the anesthesiologist had neatly arranged vials of antibiotics, narcotics, and Versed, an amnesic agent to block traumatic memories of surgery that might be registered by Matt's brain even under anesthesia.

Scrubbed, gowned, and masked, the main OR nurse unwrapped the sterile packages handed to her by the circulating nurse and carefully arranged the equipment on carts with sterile drapes. Each set of gleaming instruments was checked for all its components, down to the last titanium screw. With the main instruments arrayed in front of her, the OR nurse had a cart on either side for ancillary instruments, such as hemostats and scalpels. In a large white plastic pan lay the sterile Brain-Gate implant with wires leading to the titanium pedestal. This would be connected to a preamplifier that enlarged the signals from Matt's nerves and relayed them to the BrainGate computer system—if all went well.

Behind the carts of instruments, mounted on a wall, was a white light box where Matt's MRI scans were displayed near a computer printout that listed the x, y, and z coordinates. Gerhard would match these to the coordinates on the Leksell frame to locate the correct site of the implant before his first incision.

Outside the OR was a large sink for three people, with dispensers of surgical masks, scrub brushes, and antiseptic Avagard. At the side of the sink was a red biohazard container on wheels. Gerhard used a foot pedal to squeeze out a handful of the brown antiseptic, lathered the solution up to his elbows, and rinsed it off. Then he gently kicked open the OR door, keeping his hands sterile. He wore a lead apron because X-rays would be taken during surgery; over this garment he donned a sterile gown and two layers of sterile gloves with help from the circulating nurse. Matt had recovered from the Ativan to a great extent, and his eyes were glancing around the OR. Gerhard again briefly discussed the operation and ensured that Matt still wanted to go ahead. This was absolutely his last chance to withdraw from the study.

"Let's do it," Matt whispered.

Gerhard smiled behind his mask.

"You heard him," he told the surgical team. "Let's do it!"

Murmurs of enthusiastic approval rose from the blue-clad group.

Gerhard slipped sterile blue sleeves on the handles of the circular overhead lights so he could adjust them during surgery. Now the anesthesiologist injected Matt with sedatives and muscle relaxants that would completely paralyze his muscles and prevent spasticity. Matt was also at risk of autonomic dysreflexia—his blood pressure could rise to dangerous levels because his sympathetic nervous system was cut off from the brain's control. But the anesthesiologist felt confident about controlling this condition and placed Matt under anesthesia.

During any operation, it is essential to monitor and control the patient's position, pain, and general physiology. Matt's spinal cord injury had made him poikilothermic, like a reptile, so his body temperature was maintained by a warming blanket called the Bair Hugger. A bladder catheter drained his urine, and an intravenous catheter kept him hydrated. A Kendall pneumatic compression device squeezed his legs to prevent blood clots. The anesthesiologist had already placed EKG leads to monitor his heart, and an automatic blood pressure cuff was in place. Apollo monitored his levels of oxygen and carbon dioxide.

Matt's head was supported over the edge of the OR table, and then Gerhard locked it with the Leksell halo frame onto a clamp attached to the table. Finally, Matt was strapped onto the table to prevent movements that might interfere with the precise placement of the BrainGate. Now his head and body were fully paralyzed and immobilized. During the process of locking Matt to the operating table, one of Gerhard's gloves touched a contaminated surface, so he took it off and snapped it toward the wastebasket. It was a perfect shot. Gerhard was on top of his OR basketball game—a propitious sign for the surgery.

Gerhard checked the target coordinates obtained by the MRI and set them on the Leksell's semicircular metallic piece with measurements in degrees for rotation and millimeters from the midline. After rechecking the numbers, Gerhard attached the arc to the frame.

With these preparatory measures in place, Matt's scalp was cleaned with iodine antiseptic, and he was covered by a sterile drape. A technician wheeled in the Philips Pulsera, an X-ray machine that was inched over to Matt. Now Matt's brain was visible on a television screen, along with the Leksell stereotactic system that would guide Gerhard to the exact implant site determined by the MRI. Gerhard stared at the image, then asked for the table height to be adjusted until he felt certain of the coordinates. He was ready to begin.

After injecting Matt with anesthetic lidocaine and epinephrine to control bleeding, Gerhard made a four-inch longitudinal incision on the right side of Matt's scalp.

Once again, Matt was at knifepoint, but this time for a potentially therapeutic reason.

With a scissor instrument that had a fork on either end, Gerhard spread apart the soft tissues of the scalp. A gauze sponge slowed down the bleeding, and the neurosurgeon continued probing with the forceps.

With a periosteal elevator, Gerhard peeled back the skin to expose Matt's ivory skull. On the upper screen of the anesthesia cart, Matt's vital signs and cardiac-rhythm tracing looked strong. Gerhard marked the target area, made a half-inch burr hole with a pneumatic drill that easily cut through the centimeter-thick skull, then cut the bone with an electric saw to make a flap about two inches in diameter. He suctioned away some blood to expose the dura mater, the brain's whitish fibrous membrane of protective tissue.

"The last layer of protection from a neurosurgeon," commented the OR nurse.

"That's right." Gerhard sliced through the dura until he exposed the spidery arachnoid membrane.

"When I cut the membrane, Matt will lose some cerebrospinal fluid, and air will go in. He'll have a major headache, so we'll give him some Fioricet," he said, referring to the combination of Tylenol, caffeine, and butalbital.

Now Matt's motor cortex, with the prominent gyrus that controlled his arm and hand, was visible, the yellowish brain tissue reddened at the blood vessels. Once again, the neurosurgical team checked the coordinates on the Leksell frame; they still appeared to be accurate. It was time to place the BrainGate onto the cortex of Matt's brain.

Gerhard mounted a small pneumatic wand onto the stereotactic arc and targeted it away from large blood vessels that might bleed excessively. This device ensured a precise, fast, and nontraumatic implantation. Gerhard slowly advanced the inserter over the BrainGate implant to test and charge it. Then, with a small whoosh of air, he implanted the array at a speed of seven meters per second into Matt's brain tissue to a depth of a millimeter. As expected, there was a small amount of bleeding, which was irrigated and suctioned. When the surgical field was clearly visible, Gerhard felt that the BrainGate array was at the right position in the cortex.

Burke leaned over and whispered to Shawn, "He nailed it!"

All the steps in the surgery were essential, but placing the implant was crucial. It had to be perfect.

Now it was time to affix the cylindrical pedestal. Its base had a flat ring with eight holes for titanium screws that anchored it to Matt's skull. Gerhard used two titanium plates shaped like dog bones over the wires and secured them to the scalp with screws. With the implant fixed, Gerhard closed the dura membrane with Nurolon sutures. He applied a DuraGen patch to repair his earlier incision and provide watertight closure so that the cerebrospinal fluid bathing Matt's central nervous system would not leak out. The bone flap was fixed at three points with the titanium dog bones and small screws.

Bone has a rich blood supply, so Gerhard irrigated the area and used a cautery wand to burn the small arteries and stop any bleeding. With absorbable sutures he closed the three layers of tissue above the bone and used a sterile staple gun to finish the wound repair. Then Gerhard applied a special antiseptic dressing around the pedestal site: the Biopatch.

Surgical blood loss was only seven ounces, less than a cup. Matt's ventilator settings were the same as always. His vital signs remained stable, his pulse rising to a maximum of slightly above a hundred beats per minute.

<hr />

Adrift on anesthesia after the three-hour surgery, Matt lay still in the neurosurgical intensive care unit. His surgery had gone as smoothly as could be expected for the first human being to receive the implant. And so, nine months after helping to design the BrainGate study; getting the protocol approved by the FDA, the Rhode Island Hospital IRB, New England IRB, and Sargent Rehabilitation Center; recruiting Matt (and his family); screening him for all the study criteria; and planning his surgery, I had arrived at the moment that started this story. I stood at the bedside of a paralyzed young man with a brain implant that might help him control a computer with his thoughts. Looking at Matt, with the pedestal protruding from his skull, his shaved head covered with pure white dressings, I thought, All we need to worry about now are bleeding and infection.

After chatting briefly with my BrainGate colleagues, I greeted the Nagles at the café.

"Matt's doing fine," I said, "but he has the worst headache of his life. His meds will make it seem like a dull ache. Do you know what he did when I just saw him?"

Ellen looked at Pat and said, "No. . . ." I gathered that Matt had a habit of surprising them.

"I'm pretty sure he winked at me, right through the anesthesia." I laughed and said, "Reminded me of the first time we met—for the informed consent."

"I can't believe Mattie." Pat grinned, his face crinkling around the eyes. He shook his head. "Three years ago, we almost lost him forever. Now he's the first person in the world to have the BrainGate, and he winks at his doctor about it."

"I think he was trying to make me feel better," I said. "Matt probably knew that I was more nervous about the surgery than he was."

13

Wired, Connected, and
Waiting for Signals

Matt, June to December 2004

I returned to the ICU after talking to the Nagles and stood with my hands gripping the cold bedrails, looking at Matt as a parent looks down at a sleeping child in a crib, with wonder and fear. A nurse and I examined a small pressure ulcer on his back that could become infected and send bacteria into the bloodstream, which could in turn infect the brain implant. We agreed that he would be turned regularly from side to side.

That evening, a nurse suctioned thick green mucus from Matt's lungs, and I worried that the surgery had increased his risk of pneumonia. Although alert, he had a throbbing headache, so his intravenous Dilaudid was increased. Then nausea engulfed him, but intravenous Zofran helped settle his stomach. On occasion, Matt was offered sedation for anxiety, but he declined.

The following morning, Matt was sleepy, and he refused his lung suctioning because he felt uncomfortable. His pulse was around a hundred, and his systolic blood pressure was low. He didn't feel dizzy, though, and did not seem dehydrated. A CT scan showed no complications. I reassured Matt that, if all went well, he could soon go back to New England Sinai.

On the morning of his discharge, two days after the BrainGate implant, Matt refused the blood draw to assess for infections, bleeding, and electrolyte abnormalities. My strong-willed patient had a mind of his own. Maybe it was his nature or an adaptation to his disabled condition, or both. Perhaps the phlebotomist had failed to find a vein once or twice, and Matt simply didn't have the patience. Whatever the case, his condition appeared stable, so I chose not to fight this particular battle. As a precaution in case the FDA, New England IRB, or Cyberkinetics raised their eyebrows about my deviation from the protocol, I asked the nurse to document Matt's refusal in the medical record.

At New England Sinai, Matt slowly got back to his routine. Pain management in the first few days was a major problem, and Dr. Bloom had to increase Matt's narcotics. Because of insomnia, he required more sleeping medication than usual. Although drainage at the surgical site persisted, it didn't appear infectious and was probably an inflammatory reaction to a foreign body: the BrainGate implant. But having seen infections develop at prostheses such as knee replacements, I was acutely aware of how subtly the drainage could change, heralding a serious infection. As a result, for many nights I went to sleep wondering if an infection would develop and when we could safely start testing the device.

Matt's nurses at New England Sinai closely watched the implant site. Gerhard also made regular trips to ensure that the drainage did not become viscous or yellowish. After every visit, Gerhard sent a report for my research files. Once, in response to my query about a possible infection, he wrote, "No way," but added, "Not yet."

Gerhard was right to be confident, but he was also right to hedge his bets. In the monkey studies, implants had been removed for various reasons, including infection. We all feared having to take Matt back to the operating room, but if necessary, it would be done. The drainage remained noninfectious, but we remained uneasy. Until the seepage stopped, we could not connect the fiber-optic cable to the pedestal on Matt's head to record brain signals.

A week after Matt's surgery, I was surprised to learn that he had contacted Jeffrey White at the *Patriot Ledger* in Weymouth, which had followed his story ever since the stabbing. Matt was apparently delighted to have the implant and eager to inform the world. Perhaps he hoped to enter other research studies, to help his breathing and even repair his spinal cord via stem cells.

Burke asked me to help call off the interview, so I explained our concerns to Matt: He was our first patient. We didn't know if even a single neuron would give recordings. If we did get nerve signals, would he be able to control a cursor? We were still uncertain about the surgical outcome because of the drainage. It would be a clinical as well as a media disaster if there were a brain infection. We had no data to present. It would be wrong to raise false hope about the BrainGate. Reporters would trespass on the privacy of other patients, and they might even damage the BrainGate equipment.

Matt was unmoved and decided to go ahead with the interview.

Burke hastily arranged a teleconference with Tim, Gerhard, John, and me. We came up with a few approaches. Kari Watson, our media consultant, tried offering first rights on definitive results to the reporter—if he delayed his story. I appealed to Jeffrey as Matt's research physician (for whatever that was worth); that was also to no avail. Ellen Nagle was sympathetic but said, "Matt drives the bus."

So I got back to the bus driver for another long discussion. First, I told myself to see his point of view and not to feel frustrated. Matt agreed to not discuss his surgery; he would talk only about his current status. When the reporter called me for comments, I again pleaded with him—for the sake of the study and all people with severe disabilities—to delay the article, but I didn't feel optimistic. The next day, in a teleconference with the public-relations staff at New England Sinai, I learned that they had restricted media access to the facility because of concerns that one reporter would soon lead to others, like ants hungry for a sugar spill.

Stubbornly, the drainage continued for two weeks—then it stopped. Gerhard sent an e-mail saying that Matt's wound was healed and the area around the connector was forming healthy tissue. We could start the recording sessions, but Gerhard emphasized that the Biopatch should always be around the connector to prevent infections. In closing, Gerhard noted that Matt seemed to think that his left thumb had regained some movement. I replied that on the "day before surgery, I had reminded Matt that there were no guarantees as to how well the BrainGate would work, if at all, and that it was not designed for increased movement. This was in response to his comment that he might be able to move his arm!"

Nonetheless, Matt asked the clinical engineer, Abe Caplan, when he would be able to move his arm. Matt clearly had normal cognitive function and must have understood what the BrainGate could and could not do. But perhaps the concept of the BrainGate was so powerful that Matt *imagined* it made his thumb work. And maybe there was an element of wishful thinking.

I had to visit Matt, examine him, and provide an explanation. At New England Sinai, in the hallway to Matt's room, a ProCare device stood ready to monitor the oxygen saturation, pulse, and blood pressure in an emergency. Boxes of gloves lined the corridor rails near the patient doorways, a reminder that infections could strike at any time. A box mounted on the wall contained nasal oxygen tubing, a plastic mask, and a stainless-steel device, like the one used when Matt was stabbed, to force open the mouth and slide in a plastic trach tube. Pastel paintings of summer beaches decorated the walls.

I greeted Matt and began my examination. Not even a single muscle fiber twitched in his thumb. Once again I explained that the BrainGate implant was only a recording device, not a stimulating one. Matt was crushed. But I had to be clear about the BrainGate. If Matt perceived recovery in his hand, he would be unhappy if the movement went away. Then, he might even believe that the device had malfunctioned. Fortunately, this distraction faded, and Matt stopped asking about his hand a month after the implant surgery. He knew that eventually the BrainGate might control muscle electrodes that moved the hand. Later on, Burke told me Matt had also talked about using the BrainGate to regain his sexual function. He had appeared to understand my comments about paralysis, so I did not discuss this issue with him.

On July 12, 2004, in front of an audience that included most of his research team, Matt's head and the pedestal were cleaned with antiseptics to connect the preamplifier, which was linked to the computer system via a fiber-optic cable. Matt asked some detailed questions about the procedure, but Burke answered with generalities, as he didn't know what to expect. Having practiced the connections and optimal computer settings, the clinical engineer, Abe, was ready for a dazzling moment in neuroscience research. All we wanted from the hundred microelectrodes on the implant were recordings from six neurons. Only six.

All the components of the BrainGate system were poised to record neurons from Matt's motor cortex. Then Abe turned on the switches for the circuit between Matt, the BrainGate implant, the pedestal, the preamplifier, the fiber-optic cable, and the computer. On the occasion of the first neuronal recording from the BrainGate implant in the first human being, there was nothing.

Nothing except electrical noise.

A tangle of random electrical signals.

Static.

A shocked research team.

Excruciating disappointment.

After cruising at six hundred miles per hour and thirty thousand feet, the BrainGate study plummeted to earth with the realization that there were no usable nerve recordings from the implant. The lack of signals could be due to many reasons: infection, bleeding, incorrect placement of the sensor implant, or a host of technical problems involving the microelectrodes, the cable, the software, or the computer hardware.

John Donoghue guided the clinical engineers in troubleshooting the system. As they checked each possible problem, no signals appeared; there were only random, jagged lines that sounded like tides against a sea wall. Next, the research team reviewed the surgical video to see if anything had gone wrong in the operating room. Had a wire been cut accidentally? Had the implant shifted its position during surgical closure?

When I discussed the situation with Giselle, who was still nervous about my involvement in the BrainGate study, the news intensified her unease. But she responded with great restraint and said that it was still early. As a psychiatrist, she had extensive experience dealing with anxiety disorders, so I tried using her techniques to calm myself. I said nothing to Nita and Jacob, who had been so excited about their dad's research.

Every day I called the clinical engineers or Burke for an update. They had methodically excluded certain problems. First, they filtered out environmental noise from fluorescent lights and electrical equipment. Then the technicians checked all the connections from Matt to the computer system—and got only noise from the hospital intercom system. This was precisely the kind of situation we had feared while trying to minimize the media fanfare. Excessive optimism could lead to unabated disappointment. Then the entire project would be doomed.

In view of this potential catastrophe, there was even more pressure to recruit another patient. Public relations became the first priority. I talked to Shake-A-Leg, a recreational organization for people with severe disabilities. Leo Canuel from PARI agreed to send a mass mailing, but it would have the disclaimer that his organization did "not endorse or represent Cyberkinetics." PARI provides "training, equipment, advocacy, and counseling to help" people with disabilities; education is part of its mission. PARI also helped me coordinate a lecture in August, which was advertised in the *Providence Journal* and the *Boston Herald*. As the final step in our public-relations strategy, Cyberkinetics sent a videotape about the study to assisted-living centers.

On July 17, 2004, the *Patriot Ledger* ran the first media story ever written on Matt and the BrainGate. A cap covered the pedestal in the photograph, but some details about the BrainGate implant were revealed. The whole effort to block the publication had been intense and draining. And I felt conflicted about balancing Matt's freedom of speech with the research project's goals.

To his credit, Matt tried to undo some of the damage; he "declined to talk about the study in depth during an interview in his hospital room. Nagle's family also would not talk about the study or his operation. All Patrick Nagle would say about his son is that he has faith in science. He is willing to try anything, because it might help people."

Jeffrey White's story did provide me with new insight into Matt. He had gone to a lecture by Christopher Reeve, who described research that could "eventually eradicate paralysis" and urged people to be optimistic. After that event, Matt said to himself, "Hope. Hope. Hope," and, with the staff's help, started hunting cyberspace for the latest research. Matt focused "on stem cell research, looking into the issue and asking everyone . . . why Europe is so far ahead of the United States on stem cell studies."

At the end of the story, Matt commented: "I have total faith that I will walk again. I expect to be off the ventilator and moving my arms within two years. There's so much research being done, it's only a matter of time."

With the publication of the newspaper article, we felt even more pressure to produce results. Cyberkinetics brought in software and hardware experts with specialized diagnostic equipment and a wealth of experience in biomedical engineering. They repeatedly pored over flowcharts of all that could go wrong with the BrainGate system, first scrutinizing the most vulnerable points and then moving on to others. Tim helped us remain calm.

"From years of working with medical devices, I've learned to be patient," he said. "Let things unfold. We can figure this out."

Troubleshooting the BrainGate continued at a nervous, intense pace. Interference from the cleaning solution for the pedestal could have affected the recordings. It was also difficult to see, with skin and hair in the way, if the connector was all the way down on the pedestal. Abe was understandably afraid of pushing it too hard and damaging the surgical site. Finally, Matt occasionally had pain when the connector was screwed on to the pedestal, as if hair were getting caught inside. The research team spoke in general terms to Matt about adjusting the system to get nerve signals; I didn't express my worries that everything to this point was all for nothing.

Before the end of July 2004, Gerhard used a different cleanser to carefully swab the pedestal, and he tightly screwed on the connector that was attached to the fiber-optic cable leading to the computer. That completed the circuit. Electrical signals from individual neurons raced to the BrainGate computer, where the waves danced across the screen.

The Cyberkinetics team let out a collective shout of triumph and joy that might have been heard as far away as Wessagusset Beach, where Matt had been stabbed and the seagulls circled over the water. For now, at least, we no longer feared that the first human BrainGate implant had failed.

About thirty neurons gave consistent signals, far more than required to control the cursor. Each neuron created a little ticking sound when amplified. As Matt used the BrainGate, the tone and frequency of the nerve signals changed, sounding like rain against a window.

We could hear his brain.

To see if Matt could control the nerve signals, we asked him to imagine moving his paralyzed hand, but he could not. Then Burke played to Matt's athletic ability and asked him to close his eyes as he imagined throwing a football. Matt visualized himself as a quarterback, pivoted his trunk, drew his arm back, and threw a perfect pass to his receiver in the end zone. Now the neurons fired selectively.

Matt's first response when he moved the cursor with only his thoughts was "Holy shit!" It struck him that the BrainGate was, in some ways, a temporary escape from his paralysis.

At one of the sessions, Abe connected the pedestal to the preamplifier, which sent signals via a fiber-optic cable to the BrainGate computer. He showed Matt a reproducible and stable recording from a single neuron and asked him to modulate this signal. Matt was able to change the nature of the signal by merely thinking about it. In all, there were about thirty neuronal tracings, an encouraging sign. But what had happened to the other seventy microelectrode signals? Was there some bleeding or inflammation around the implant? Were the signals too weak? Was the implant not in the proper position? Or was there no rational explanation? (Scientists are, unfortunately, not allowed to say that.)

Abe continued gathering data that correlated electrical patterns with Matt's intentions to move the cursor on the computer screen. At first, Matt imagined opening and closing his hand; three neurons became active with hand closure (see figure 13.1).

While Matt imagined moving his hand in specific directions, the BrainGate system recorded the associated electrical patterns. Later, when Matt wanted to move the computer's cursor, this information was used to interpret the electrical patterns and control the cursor with only his thoughts. We could read his mind.

Soon, Matt began to consistently control the cursor. As Abe tried different versions of the software, Matt occasionally felt that he had lost some control and took it to heart. "I'm just not doing as well as I was last week."

"We're experimenting with how to get the best results," Abe said.

"It's not just you," Burke said to Matt. "You can't take all of this on yourself."

But Matt's competitive spirit as an athlete surfaced. "I've got to do better today."

Some days, he was tired and drowsy, especially after pain medications, and fell asleep even when connected to the BrainGate. Abe would wait quietly for an hour or more while Matt had a nap.

Once Matt developed sufficient control, he accurately performed 399 out of 480 trials by starting with the cursor in the center of the screen, then going out as directed (north, south, east, and west). Abe gave him progressively more challenging tasks. He asked Matt to move the cursor—with only his thoughts—to one of four boxes that randomly became a colored target on the computer screen. When he did this task perfectly, the number of boxes increased and the task became more difficult (see figure 13.2).

With Abe's guidance, Matt trained his brain to move the cursor with only his thoughts, as if he were capable of telekinesis. Within a month, he achieved reliable and consistent control of the cursor.

A while after this mastery of the BrainGate, Matt and I had a discussion in which we tried to understand each other and to prepare for a long and challenging season. I described my experiences in treating people with tetraplegia, to let Matt know that I could imagine his damaged life and sympathize. Then Matt told his story.

"Lots of tears, Doc. Lots of tears."

Matt sighed. His odyssey had been interrupted by quicksands of depression.

"One day I'm OK, the next I want to go to the terminal wing. A priest told me God uses some people to show others what life is like. I've accepted that. My relationship with God has gotten stronger and stronger. There's a plan for me. You know, you think you're the only one, but then you hear support groups talking. I hope to inspire people with the BrainGate. I've got to go forward from here. Can't quit. Once I told my mother I wanted to die, and she said, 'Then I'll go with you.' That's love, that's real love."

Matt paused, his emotions choking off his words.

"I've got to go on for my friends. Twenty friends were at the beach that night. Only four have stuck around. I know who my friends are, and I love them to death."

Matt continued, expressing his gratitude to the staff at New England Sinai for their caring and their patience. "Hospitals are hospitals, but the people here have been very good."

Matt's voice was more hoarse than usual, so I held up a cup with a straw. He nodded and took a long sip while the ventilator kept on puffing. Having poured out his emotions, he was quiet.

After a few days of training and data collection to refine his control of the cursor, Matt graduated to the video game Pong and was able to move a bar (like a Ping-Pong paddle) across the bottom of the screen (the table) to block the incoming ball. Of course, at first there were a number of missed balls, but eventually Matt blocked the ball in about two out of three trials.

A more complicated video game involved moving the cursor to money bags before a masked bandit reached them. Here, Matt had a little more difficulty because the targets moved faster. Eventually, he wanted to play Nintendo games with the BrainGate.

During the fall of 2004, Matt had some tough days. There had been so much anticipation and anxiety that every moment Matt could spend with the BrainGate was precious. As a night person, he wasn't prepared for the intensity of the morning sessions with the Cyberkinetics technicians, and sometimes he dozed off. Abe Caplan realized that it was best to shift his own schedule to fit Matt's and started the recordings after one o'clock in the afternoon. They got into a routine and a rhythm that worked for both. They spent some time with the BrainGate, chatted for a while as a break, and went back to the study before Abe left.

One of the BrainGate experiments coupled the neural cursor to a television via infrared technology similar to that used in remote-control devices. On the computer screen were boxes labeled POWER, CHANNEL + and –, and VOLUME + and – (see figure 13.3).

"Matt, I want you to move the cursor to the power box," Abe said, after placing the cursor in a corner of the screen.

Matt concentrated, imagining how his hand would move the cursor to the box. The arrow wavered, started to move, then landed on the box. The television switched on.

"Great job," Abe said. "Now I want you to turn up the volume."

Matt nodded and stared at the screen. The cursor danced over to the volume box, and the sound became louder.

"How about changing the channel?" Matt whispered. "I think the Red Sox are on."

Abe smiled, and Matt clicked up the channel box to ESPN. With practice, Matt used the BrainGate cursor to select boxes on the computer screen, allowing him to turn his television on and off, select channels, and change the volume.

A major purpose of the BrainGate was to improve the access of people with disabilities to the outside world. Once Matt controlled the cursor, he opened and closed e-mails via the BrainGate. It was exciting to see Matt escape his small hospital room into endless cyberspace. But he was not yet able to accurately type with the BrainGate to respond to e-mails.

Remarkably, with all these tasks, Matt stated precisely what he was going to do, proving his voluntary control. In other words, the cursor didn't randomly wander over to the icon for opening an e-mail.

Once Matt had mastered more complicated maneuvers with the cursor, his next challenge was to use Microsoft Paint software. He approximated squares and circles, but even after a great deal of training, he lacked accuracy. With concentration and coaching, Matt drew a "circle" that appeared to be more of an oval—but for a paralyzed patient using only his thoughts, it was a work of art. He started off at the lower-left corner of the screen with a paint bucket of black ink and used the cursor to go near all the other corners and back. Encouraged by his progress, I hoped that the Paint software might allow Matt to produce some art; later on, I asked Abe if he could use different colors to give Matt a creative outlet.

By the end of August, Matt became well attuned to the system, carrying on a conversation or whistling as he moved the cursor. Eventually he controlled a prosthetic hand that was hooked up to the computer. By raising the cursor to the top of the screen, he opened the hand, and by moving the cursor downward, he closed the hand (see figure 13.4). Matt achieved this feat after only a few trials—while looking at the hand and not the cursor. For another victory over paralysis, Matt used a multi-jointed robotic limb to grasp and transport objects, such as a ball.

In early September, Burke asked the research team about releasing Matt's dramatic results at the American Academy of Physical Medicine and Rehabilitation (AAPM&R) meeting. We considered a few factors. No one else had enrolled in the study, but more people would want the BrainGate once they heard about Matt. Tim would be much more successful at raising funds. For a biomedical start-up, cash is always a problem because of research and development costs (such as computer equipment, software, and consultants). A related matter was the upcoming initial public offering (IPO) for stock in Cyberkinetics. Matt's incredible performances would impress venture capitalists and other investors. Most important, as the premier meeting in the field of rehabilitation medicine, the AAPM&R event would have media coverage.

We agreed to present Matt's preliminary data on a poster that would show the BrainGate implant, the computer screen with neuronal recordings, and Matt controlling a cursor and a television—with his thoughts.

Burke showcased Matt's results on a CD. First, a computer simulation showed a patient using the BrainGate to control his environment. Then Matt performed all his computer tasks, one after another. I smiled at Burke and savored the moment. I had heard about and seen these performances, but never on one CD for a public presentation.

On September 30, Burke e-mailed the poster for the AAPM&R meeting. Sequential frames showed the theoretical basis for the BrainGate, inclusion criteria, details of the surgery, electrical signals from individual neurons, Matt using the BrainGate to control a television, and a schematic figure of potential applications of the BrainGate. Explanatory text described the protocol, the neural recordings, and Matt's abilities with the BrainGate. It was a work of art.

On recovering from the initial feeling of accomplishment, I critically examined the poster. I corrected a few typos, clarified some technical language, added surgical details, described my examinations, and made the data more meaningful to clinicians. After comments from my coauthors, especially John Donoghue, we produced a final poster that was enlarged to four by six feet.[1]

On another vital front, Cyberkinetics went public to raise more money for the BrainGate study. To avoid the heavy fees of an investment banker, Tim used the shell of a defunct company still listed on the NASDAQ stock exchange, then changed the name to Cyberkinetics after the stock started trading.

Shortly before the IPO, Burke asked me to speak to a potential investor from a medical hedge fund. Armed with a doctorate in biomedical engineering, the investor posed a series of questions: What exactly was the technology? What was the epidemiology of spinal cord injury? What percent of people with SCI would qualify for the BrainGate? Would people with disabilities accept the device? Would insurance companies provide reimbursement? What regulatory hurdles did I anticipate?

At times, the conversation became uncomfortable as the investor tried to pin down my answers. With millions of dollars to invest, he wanted clear and detailed answers with minimal uncertainty. This was one of a few occasions in which there was a conflict between serving as a clinical scientist, a principal investigator paid by Cyberkinetics, and a spokesperson for the study. Burke was also in on this conference call, and I was as honest as possible while keeping in mind the goals of Cyberkinetics for fund-raising.

I had declined an offer to buy CYKN stock after considering the ethical issues and financial risks. Cyberkinetics started trading at $2.50 per share, but after Matt's achievement with the BrainGate was reported internationally, the stock almost tripled. Within a month, however, CYKN stock started drifting down.

Burke had told Matt about presenting his data at the national rehabilitation meeting, and he seemed agreeable. But once Matt realized he was making scientific history, he said, "If you're going to use my data, put my name on the poster."

"This is a medical meeting," Burke said. "We don't do that."

"What do you mean?"

Burke felt that Matt was being a little difficult. Having managed a hundred clinical trials, he had never dealt with such a request.

They stared at each other. Burke blinked.

"Matt, what are you really trying to get out of this?" he asked.

"The clock's ticking," Matt said, and he looked out the window. "I'm going to die, man. I'm going to die. I'm a quad. It's a matter of time. I want my name out there so people come to me, I'll get into other studies."

Burke called me to discuss this dilemma, and the research team had another teleconference with our public-relations consultant, Kari Watson. She was concerned that announcing Matt's name could "diminish the scientific credibility of what has been accomplished and, sadly, put the patient in the position of performing tricks" for the media. Also, releasing his name the day after the Cyberkinetics IPO would look like a "ploy to pump the stock." As a compromise, Kari recommended disclosing

Matt's name only in the news release for the AAPM&R meeting. Sargent Rehabilitation Center should arrange a press briefing to introduce Matt and discuss the results. She felt this would give recognition to both Matt and Cyberkinetics in a professional manner.

For this study, we had one patient, the FDA, Sargent Rehabilitation Center, and three research boards (Rhode Island Hospital, New England IRB, and New England Sinai). A lot of people were watching how we conducted the research. We agreed to include Matt's name in the press release but not the poster, and he was happy with this solution. Even though I was violating a research subject's confidentiality at his own request, I obtained a release from Matt for ethical and regulatory reasons.

On October 5, 2004, Kari sent an urgent e-mail that Kevin Maney at *USA Today* wanted an interview the next day. She suggested that I offer clinical remarks with "appropriate caveats about the preliminary nature of the results." The next day, I received the news release in which Sargent Rehabilitation Center announced that I would present preliminary clinical results of the BrainGate study.

"I am feeling well," Matt said in the release, "and am using the BrainGate about twice a week. In the past few weeks, I have been able to control my TV using a computer cursor that I can move with my thoughts. I hope that someday I will be able to do even more with the BrainGate, like move my arms with a muscle stimulation device. The BrainGate has changed my life."

At the AAPM&R meeting in Phoenix, Arizona, most of the conference goers were impressed, but a few simply didn't believe Matt had controlled the cursor with an accuracy that was far better than simulated random activity. Then they saw the videos of Matt's brain signals when he imagined closing his hand, of Matt playing the video game, and of Matt with e-mail. Finally, Matt turned his television on and off, changed channels, and adjusted the volume—after announcing his intentions. Then the skeptics became believers.

When I returned from Phoenix, I was shocked to learn of Christopher Reeve's death on October 10, 2004. An inspiring and tireless advocate for disability rights and research, Reeve had reportedly succumbed to an infected pressure ulcer—a bedsore. In this day and age, with the level of care he was receiving, it was unthinkable that Reeve could die of such a complication. To prevent pressure ulcers in people with spinal cord injury, who often have paralysis and sensory loss, it is essential to reposition them frequently to avoid tissue damage from pressure.

People magazine reported that, the night before his death, Christopher Reeve had watched his son's hockey game, been driven back home, and then watched the Yankees-Twins game. During this busy day, perhaps a pressure ulcer worsened, became infected, and caused widespread infection (sepsis) and multisystem failure. Christopher Reeve's heart failed and his lungs drowned, in spite of all his courage and optimism.

A *USA Today* story entitled "Scientists Gingerly Tap into Brain's Power" was illustrated with a cross section of a brain implanted with a device that radiated information. After mentioning Matt's ability with video games, the reporter added that the BrainGate study had "far-reaching implications beyond Pong." On the ethical issues, Arthur Caplan at the University of Pennsylvania commented, "There are those who say that this is slippery slope stuff, that this technology is opening the door to dangerous technologies that could enhance, improve and optimize someone. But I am unwilling to hold hostage this kind of exciting medical research for those kinds of fears."

Miguel Nicolelis, a researcher at Duke University, gave a contrasting response: "I am very skeptical. They seem to want to simply push their views and make a buck without much consideration of what is appropriate and safe to suggest to different patients." In the July 2004 issue of *Neuroscience*, the Duke team had described monkeys with implants that allowed them to move a robot arm by only thinking about it.

Burke was pleased with our publicity but wanted a different slant, he said in an e-mail to the research team: "It would be very nice to focus in the upcoming interviews on what we hope to do with the system in the *future*, such as better links to computer-based devices," like environmental control systems, Internet use, and computerized muscle stimulation of paralyzed limbs. "The *Globe* and *USA Today* articles were nice. However, they show how reporters like to focus on the games Patient One has been able to perform, and not the potential in increasing independence. I'm somewhat concerned we may come across to the casual reader as an expensive Game Boy."

For a small company like Cyberkinetics, publicity could make or break the project—and the corporation. In the biomedical world, a war chest of $9 million was a pittance when it came to supporting the research and development for a first human trial. Burke realized, however, that we "often have little control over these things when giving an interview."

Later that day, Kari Watson sent the research group a list of interested reporters. In a phone interview with Manfred Dworschak from *Der Spiegel*, I discussed the BrainGate at its current stage and a future wireless device to control robots. While waiting for a bus on my way to give a BrainGate lecture at Tufts University, I talked with Roxanne Khamsi from Nature.com. In the *Providence Journal*, Felice Freyer described Matt as a "man who can't move his arms and legs, [but] has been checking his e-mail and changing television channels—just by thinking about it."

All of this media attention was exciting, but Burke was already looking ahead. In an e-mail with the subject line "Next Generation BrainGate Device," he wrote that Cyberkinetics hoped to develop "a useful medical device" for people with severe disabilities within the next three years. For 2005, Burke envisioned software that "could be used by the patient every day with the help of a caregiver" and no technician. Burke wanted advice on how to improve the "speed, accuracy, and ease of use" of the new BrainGate system. New devices would have to "work through the PC with all major environmental control units, Internet access, and word processing." For 2006, Cyberkinetics planned a wireless device with a "smaller computer that was powered off the wheelchair." This might allow the patient to send out nerve signals from the brain and control a computer. In 2007 and beyond, Burke hoped for "advanced versions that would use/link with robotics, computerized muscle stimulation, etc."

||||||||||||||||||||||||||||||||||||

Emily Gold Boutilier at the *Brown Alumni Magazine* and other reporters wanted to interview Matt. To avoid multiple trips, I arranged a media event at Sargent Rehabilitation Center and gave out a press kit that included the CD of Matt using the BrainGate. Ellen drove him down in the Nagles' van, which had been modified for an electric wheelchair. I imagined that she wanted to protect Matt from sensitive questions by inquisitive reporters. She also wanted the CD. It would be carefully archived, along with numerous photos and news stories about Matt, who had become her life's mission.

Boutilier had questions about the criteria for patient selection, ethical issues, and my concerns about the surgery. She asked why we selected Matt and wanted my perspective as a rehabilitation physician. There was some discussion of Matt's personal life and a question about a girlfriend who left him soon after the injury.

"She wanted to leave, and I said I was all set," Matt said with no emotion.

Felice Freyer asked why Matt was willing to risk brain surgery to have the BrainGate. Matt said that he was quadriplegic, and there was nothing going on in his life, nothing he could look forward to. He discussed his treatment at Shepherd Rehabilitation Center and recalled three doctors who said he would never walk or breathe on his own again.

"They don't have a right to tell people things like that," Matt said. He had decided that he would show them. Matt was enthusiastic about the BrainGate because he had used it to control a prosthetic hand. One day, he hoped that the BrainGate would help him walk.

During the interview, I learned that the Channel 12 TV crew was not coming because the Boston Red Sox were on the brink of winning the 2004 World Series, ending a drought of eighty-six years. A devoted citizen of Red Sox Nation, Matt was happy to be given second billing. Besides, he was content with the public recognition of his contribution to disability research, especially when he was interviewed by *Playboy*.

While standing at a window after the media event, I saw that Ellen couldn't anchor Matt's wheelchair in the van. I went out to help clamp down the electric wheelchair and promised Ellen that future interviews would be held at Matt's facility.

All the news coverage was exciting and also entertaining. *Inside Edition*, CNN, and *20/20* called about possible interviews. Matt was thrilled about the idea of going to New York. I joked that we should go together and have a good time, but the logistics were difficult. We never got our big-city trip. Although Matt was disappointed, I was relieved. Too many media activities could affect my family time and my responsibilities as the medical director of an inpatient rehabilitation center.

On another front for treatments of spinal cord injuries, the year 2004 was especially encouraging for stem cells. Japanese scientists led by Masayoshi Ohta infused adult bone marrow stem cells into the cerebrospinal fluid circulating around the injured spinal cords of rats. Compared with untreated rats, the treated ones had less nerve tissue damage and better movement. Ohta and his team felt that marrow stem cells could "be used for autologous transplantation," in which stem cells would come from the patient's own marrow, minimizing the risk of immune rejection. These results were "promising for the clinical use of bone marrow stem cells in spinal cord injury treatment."[2]

Similarly, researchers at Ohio State University led by Daniel Ankeny transplanted adult bone marrow cells into rats with spinal cord injuries. Treated rats showed spontaneous stepping movements of the hind limbs. Bone marrow stem cells appeared to help preserve the damaged cord tissues and white matter (myelin). Also, nerve regrowth was better in the rats treated with stem cells. Tissue analysis showed that the majority of nerve outgrowths in the treated rats were parallel to the cord, as if trying to bridge the gap at the site of injury.[3]

South Korean researchers triumphantly announced that Hwang Mi-Soon, a young woman paralyzed due to an accident in 1985, was treated with stem cells from an umbilical cord. Animal studies with these cells showed that they transformed into nerve cells and improved neurological function. Forty days later, Hwang Mi-Soon reportedly regained sensation and movement in the legs, allowing her to walk with physical therapy and a walker. CT and MRI scans were consistent with neurologic recovery, showing regeneration of this woman's injured spinal cord.[4]

This extraordinary news about treating a human spinal cord injury with umbilical cord stem cells—scientifically valid or not—effectively stopped enrollment in the BrainGate study. In spite of Matt's achievements and international publicity, people with tetraplegia did not clamor for the BrainGate implant. Fervently hoping that damaged spinal cords could be repaired by stem cells even decades later, many people with severe disabilities appeared to lose interest in the BrainGate. Others preferred to wait for the miniaturized, wireless version that Cyberkinetics hoped to develop.

Undaunted, Cyberkinetics persisted with its publicity of Matt's success. Kari Watson scored a touchdown by arranging an interview with

Discoveries and Breakthroughs Inside Science, a news program of the American Institute of Physics that is syndicated across America. She e-mailed me a list of sample questions and suggestions to make the interview appear more natural. For instance, the videographers wanted shots of Matt talking to me and controlling the lights. This carefully orchestrated production was all for a ninety-second segment.

When the videographer asked about the BrainGate's history, I talked about the materials and technologies of the monkey studies. Computer hardware had become much faster, and the processing speeds had improved. In college I had used paper cards with holes to run a room-sized computer; in medical school I used a desktop Apple IIe; and now the BrainGate computers had extremely fast Intel processors (two gigahertz). Burke was concerned about revealing certain technical details, so I discussed only public material. Now that Cyberkinetics was a public company under the oversight of the Securities and Exchange Commission, I used terms like "someday" to avoid creating any illusions about the device.

Most reporters understood the benefits of our technology, but a few were unrealistic. One noted the BrainGate's potential for disabled people and went on to say that "the able-bodied stand to benefit too: [neurosurgically] implanted chips may one day control common household electronics such as TVs, lights and robotic vacuum cleaners." This conjured up an image of a man in a recliner, a case of beer and a large bowl of chips by his side, intently watching a football game while he controlled a robotic vacuum cleaner—all with his thoughts.

Matt's experience also drew attention from a British media company, which e-mailed a request to follow a new BrainGate patient from start to finish. I declined due to confidentiality and ethical regulations and instead offered an interview in London that could be syndicated to British television stations. I also corresponded with some neurosurgical colleagues at University College London, and they invited me to give a presentation on the BrainGate. Near the end of November, the media company said that the BBC was interested—if it had exclusive access to film Matt's story. Obviously, I could not authorize this, and Cyberkinetics did not. In the meantime, I finalized plans for my lecture at University College. Ironically, although the BrainGate study did not lead to New York, it did send me across the ocean to London. Unfortunately, Matt could not come along.

Shortly before leaving for England, I learned that impaired sleep and pain medications were affecting Matt's sessions with the BrainGate. Some days he did fine, but at other times he was tired and drowsy. Matt suffered with chronic pain—both physical and psychic—and as a result his sleep cycle was disrupted. Sometimes Abe was upbeat about the testing sessions, but on occasion he was frustrated because Matt just wasn't awake. Burke asked me to follow up with Dr. Bloom, who did his best to adjust Matt's sleep medications.

I went to see Matt and emphasized to him the importance of a normal sleep pattern, but that was the best I could do. On my way out, I noticed that the snow melting from Matt's window had formed a six-inch icicle, a crystalline stiletto that would pierce the earth when it fell. Before leaving the hospital grounds, I broke off the icicle and tossed it into a cedar bush.

Near the end of 2004, John Donoghue received *Discover* magazine's Innovations Award for Neuroscience for developing the BrainGate. Ruth Simmons, the president of Brown University, stated that this innovative research, "in the true spirit of the Brown curriculum, has involved many undergraduate and graduate students, as well as Professor Donoghue's faculty colleagues from various disciplines."

From my hotel in London, I went to the tube station near Queen Square, then to the large brick neurosurgery building (formerly the residence of Sir Charles Symonds). Dressed in casual clothes because of the holiday break, a group of neurologists and neurosurgeons greeted me. When I finished my presentation of Matt using the BrainGate, they quizzed me in a polite, British academic manner over cups of tea. There were some fascinating and thought-provoking questions: How long could the implant safely stay in Matt's brain? Was there a training effect with Matt's use of the system? What were the advantages over a voice-activated system?

After returning from England, I visited Matt, and we chatted for a while about my trip.

Then he became serious and said, "Doc, I want to talk about electrodes."

"What do you mean, Matt?"

"You know how I moved the robotic arm by thinking about it?"

"Yeah."

"How about using the electrodes so I can actually move my own hand?"

"Well, that's a change in the protocol, so the FDA and New England IRB would have to approve it."

"Why?"

"They oversee this project and make sure that I'm ethical and doing everything according to good clinical practice. Researchers might take advantage of people like you, who are willing to do almost anything to get better."

"OK, so what can we do about this? Can't we implant the electrodes?"

"I'll talk to Dr. Donoghue about it," I said. "We'll do our best to help you. You know, one of our goals is to use electrical stimulation with the BrainGate. That *might* help you move the arm with only the brain's electrical activity. There's no guarantee."

"I've been good for Cyberkinetics, and I know you've been good to me." Matt gave a pleading smile. "Whatever you can do . . ."

I gathered that Matt had been doing Internet research about functional electrical stimulation (FES), in which implanted or superficial electrodes stimulate muscles to move. He knew about the Cleveland FES Center and patients such as Jennifer French, the executive director of Neurotech Network,[5] who had paralysis due to a snowboarding accident that injured her spinal cord. Refusing to accept her condition, in 1999 she became the first woman to receive an FES prosthesis to help her move and transfer from one position to another, such as seated to standing. While discussing her experience with me in February 2011, French described surgery in which coiled wire electrodes were carefully placed in her paralyzed quadriceps, hamstrings, gluteus maximus, and back muscles. Her surgeon tunneled insulated wires from each electrode below her skin and connected them to a small receiver/stimulator that was surgically implanted at her abdomen. An external control box sent radio-wave commands to the receiver/stimulator, which sent currents via the electrodes to stimulate her muscles. This complex process, coordinated by a computer program, moved her legs and stabilized her back. The control unit had a green button to start stimulation, a blue one to select programs for muscle groups, and a red one to stop the FES. The book-sized device also had warning lights for a low battery and loose cable connections (see figure 13.5).

At first, French was restricted in her activities so that she could heal. Then she had an intense eight-week therapeutic exercise program to retrain her muscles for standing. During these exercise sessions, she activated the electrodes to stimulate her legs and back muscles. Then she placed her walker in front, moved forward in her chair, and hit the green button to activate the eight electrodes implanted in her legs and back muscles. A computer program slowly increased the current to "tell" her muscles to help her stand.

"It was wonderful to stand up with my own muscles, without braces," French said. "It was great to see that empty wheelchair."

Then she described walking down the aisle for her wedding. FES helped French to stand, move from one position to another, and walk for short distances. But unlike Matt, whose arms were fully paralyzed, she could use a walker with her upper extremities.

Ideally, FES systems have sensors that send electronic or mechanical information to the controller, which adjusts the stimulation. For instance, sensors can be placed in braces to detect the angle of the knee joint. When the knee bends as the limb moves forward, the sensor sends a signal to the controller, which tells the stimulator to make the quadriceps straighten the knee and continue forward movement. As in the Tibion Bionic Leg used by Linda Holmes, feedback is essential for efficient walking.

In addition to psychological well-being, French felt that standing and walking with the FES prevented some of the complications of tetraplegia, such as urinary tract infections, spasticity, osteoporosis, and pressure sores. An avid sailor before her snowboarding accident, French resumed sailing at a wheelchair-accessible facility in Florida—she loves to stand up when the water and wind are calm—and has won a gold medal in the US Disabled Sailing Championship. I hoped that someday Jennifer French would sail up to Newport, not far from New England Sinai, and Matt and I would greet her on the dock as if she had just won the World Cup (see figure 13.6).

Darrell Hover, a patient with a spinal cord injury due to an all-terrain vehicle accident, described having limited use of his hands. After his 1996 surgery for an implanted FES system and therapy to retrain his hand, he was able to use an electric razor, brush his teeth, pick up a glass, eat sandwiches without crushing them, and extend his index finger for typing.[6] Because Matt's arms were fully paralyzed, FES electrodes would have to simultaneously control his shoulder, elbow, wrist, and hand movements. People with Matt's level of injury have received FES systems (at the Cleveland FES Center), but the engineering and practical challenges are daunting. For instance, eating a bowl of cereal, with a spoon strapped to the hand, could take up to an hour. In theory, someday Matt could send nerve signals via the BrainGate system to control electrodes implanted in his arm, but it would be a complex and laborious process.[7]

When Matt and I had another discussion about FES electrodes, I noticed that his left hand was pale and puffy due to lack of movement. I did not believe he would benefit much from functional electrical stimulation. I doubted that his hand would ever again hurl a football in a perfect spiral, raise a hockey stick to score a goal, or trace the curve of a woman's face down her neck to her breast. But I held his paralyzed and insensate hand, nodded in agreement, and forced out a smile that said it was only a matter of time.

14

Skating Away from a Stroke

Garrett Mendez

G arrett Mendez started skating four years after he learned to walk. He grew up on the ice and went on to play forward for his hockey team at Notre Dame High School in Connecticut. His job was to score goals. Like Matt Nagle, he excelled at gliding between the defensemen, controlling the puck as if it were tied to his stick by a string, until a slap shot sent the small black disk flying into the net.

As a freshman at Western New England College in Massachusetts, the weekend before Thanksgiving 2005, Garrett was in an especially competitive hockey game, with more than the usual shoving and checking into the boards. Garrett skated hard to score a goal and to assist with defensive plays, rallying around his goalie. So when an opponent sent a puck spinning toward his goal, Garrett tried to block it, lunging and sliding on the ice with such force that he shot head first into the boards.

Up in the stands, Eileen and Gary Mendez gasped. Garrett's parents had cheered on their son at many games—and cringed at some hits—but this was the worst they had ever seen. In fact, until Garrett moved his legs and skated off the ice, Eileen thought he had broken his neck and was paralyzed. After the trainer checked him for a concussion and other injuries, Garrett finished the game. His neck was sore for two days, but he shrugged it off and played in a game the day after Thanksgiving, performing at his usual high level.

After the game, Garrett's legs felt weak. He had a headache, and his neck remained painful; but he didn't think much of it. Later that night, he vomited, and his face felt numb. He and his parents thought he had the flu, so he went to bed.

In the morning, Eileen found Garrett unable to speak or move. The emergency room doctor treated him for encephalitis with high-dose steroids. A radiologist saw his MRI and diagnosed a stroke in the brain stem, but the neurologist disagreed. When Garrett did not improve as expected, almost two days later, he had an MRI of the brain's arteries. It showed that Garrett had torn an artery in his neck when he crashed into the boards. As it bled and formed a clot, the blood supply to Garrett's brain stem was shut off, causing a stroke.

As a result of the brain stem stroke, Garrett could not move, communicate, eat, or breathe. Sometimes he cried out or moaned, making feral sounds. During the first week in the intensive care unit, he was severely disabled, like Matt. The control signals flowing down from the brain's surface had been blocked, damaged by the lack of blood flow. But he was fully aware of his surroundings and able to comprehend what had happened, for his higher-level functions in the brain cortex were intact. He had the locked-in syndrome.

While Garrett struggled on a ventilator, his parents—just like the Nagles four years earlier—did not know if he would survive. Garrett did, thanks to his superb physical condition. His heart and lungs could push through an entire grueling hockey game—or a brain stem stroke. He could have succumbed to complications such as infections, collapsed lungs, and abnormal heart rhythms, but he did not. Just like Matt. One of his parents or his sister, Jenn, always stayed at his side so that he would not be alone. Like most ICU patients, Garrett slept poorly and was often restless at night. In the fluorescent light of the ICU, day and night were indistinguishable. The entire Mendez family was exhausted.

The brain stem is only about three-quarters of an inch in diameter, so a stroke in this area is catastrophic. Sensory and motor nerve fibers pass through the brain stem, sending and receiving messages between the brain and the spinal cord. This vital structure regulates the heart, lungs, and blood pressure. Another crucial function of the brain stem is to form interconnections between parts of the brain, such as the cerebral cortex and the center for coordination, the cerebellum. Finally, most of the twelve cranial nerves arise from the brain stem and serve functions including vision, hearing, balance, taste, and swallowing.[1]

Garrett's parents agreed to a feeding tube, which was a turning point—now they felt he would survive. They also knew there was no miracle cure. Early on, it was unclear if Garrett would remain locked inside his brain; he was unable to move or speak but was aware of his condition. Eileen tried to keep Garrett connected to his family by reciting *Where the Wild Things Are*, his favorite book as a little boy. She wanted to give him something familiar, something to hold on to.

At this stage, he could have lived the rest of his life like Jean-Dominique Bauby, the author of *The Diving Bell and the Butterfly*. In his short, intense memoir, Bauby describes people with locked-in syndrome as "broken-winged birds, voiceless parrots, ravens of doom, who have made our nest in a dead-end corridor of the neurology department." He is also aware of the "uneasiness we cause as, rigid and mute, we make our way through a group of more fortunate patients."[2] If Garrett remained locked in, he would be more disabled than Matt. If Garrett remained locked in, he would need the BrainGate even more than Matt.

Fortunately, Garrett improved. His second week in the ICU, he moved his eyes to indicate "yes" or "no" to simple questions. His heart rate, breathing, and body temperature were more stable, and he started moving his left side. The Mendez family began to hope and started thinking about rehabilitation.

When transferred to Gaylord Hospital three weeks later, Garrett still couldn't see clearly, speak, or swallow. Completely paralyzed, he had hyperactive tone in his right arm due to the loss of his brain's input, so his elbow was bent, his hand clenched in a fist under his chin. But his brain stem had recovered enough that he was able to escape the ventilator and breathe on his own.

Garrett cried his entire first night, which he spent at the nurse's station in his wheelchair. (Later, the Mendez family was saddened to learn that Garrett thought he had been left at Gaylord for the rest of his life because he wasn't going to recover.) Eileen sped to the hospital in the morning and held him while they cried together, oblivious to a group of patients at breakfast. Then the Mendez family took turns staying in his room at night, sleeping in a folding bed next to Garrett.

<hr />

In May 2011, almost six years after his stroke, Garrett and Eileen told me about his rehabilitation. A slim, muscular young man, Garrett kept his right hand in his pocket.

"I set myself a goal the day I was wheeled in," he said. "I was going to walk out of the hospital."

"Early on," Eileen said, "Garrett showed us what he was capable of. He seemed to want a full recovery. And so did we." As she spoke, the waves of her light brown hair emphasized her point.

His therapists and nurses were also optimistic. Perhaps it was the way he communicated with his intense brown eyes and did his best. If the occupational therapist asked him to bend his right elbow once, he tried twice, straining to activate any neurons in his damaged brain. If his physical therapist had him pull up his foot, he kept trying—even though there was no movement. If the speech and swallow therapist asked him to repeat a vowel, even if he couldn't produce the sound, Garrett silently forced it out. As an athlete, Garrett knew that repetitions with weight lifting and skating drills with his teammates were essential to ultimately scoring a goal. He approached his stroke rehabilitation in the same way.

His physical therapist, Erica, started with basic neurorehabilitation: she helped Garrett hold up his head, stabilize his trunk, and control his left leg movements. His occupational therapist, Stacy Melillo, tried to restore his vision so that he could do basic daily activities. An eye-patch prevented double vision, and later she did tracking exercises to help Garrett move his right eye. For his clenched right hand, she used a splint. To progressively straighten out the tight elbow, Stacy started serial casting—an uncomfortable but necessary process. And she started retraining his left arm so that he could become left-handed, perhaps to point out his wishes or to use a remote control. His speech therapist, Kris, taught Garrett how to control his tongue and lips, how to loosen his jaw, and how to swallow again. Her first priority was his nourishment. Then she planned to help Garrett regain his speech.

Eileen did not know how even these skilled clinicians were ever going to get her son to control his body again. She told me that long after his therapists left for the day, Garrett continued his rehabilitation. His locked-in state limited his interactions with other patients and the nurses on the second shift, so he made the most of this solitude. Imagining himself back on the ice—his long-term goal—Garrett tried to move each muscle group, one at a time. As an athlete, he had experienced and intuitively understood neuroplasticity. Long hours of skating, weight lifting to strengthen his lower extremities and trunk, and innumerable drills on the ice had modified the nerve pathways in his brain. His hockey skills had developed over more than a decade; Garrett knew that he had just started the demanding process of stroke rehabilitation and many years lay ahead.

In mid-December, Jenn brought one of Garrett's favorite movies, a comedy. She was delighted by his frequent laughter—at the appropriate times. Now the Mendez family knew that, in spite of being locked-in, Garrett was aware and reacting. He was still the same Garrett.

After Garrett's daytime therapies, his parents and sister, Jenn, offered their own therapies to help Garrett with his rehabilitation homework. Looking back on Garrett's condition soon after the stroke, Eileen told me, "It didn't affect his cognition, but it was physically devastating." Her hazel eyes closed briefly with the emotion of remembering her son's helpless state.

For some reason, his left side recovered much faster than his right. Every new movement, even a flicker of his left elbow or wrist, was an occasion for celebration. Forced to become left-handed, Garrett created a new identity. Stacy retrained him to use his left arm for bathing, dressing, and toileting.

To understand the challenge of this process, a person might try keeping the dominant hand in a pocket while using the nondominant one for activities such as brushing his or her teeth, dressing, and eating. Doing these and similar tasks is even more complicated with a weak nondominant arm, but that is what Garrett did to regain his independence and dignity. As an athlete, he knew what was required to train his body. He knew that proper form—for the most effective patterns of movement—was essential when learning a new movement or exercise. He knew how to focus on the basic elements of a drill that would eventually send the puck flying into the goal. He knew how many repetitions of a drill were necessary to carry out a play and score a goal.

Garrett retrained his left arm and hand. His right arm remained paralyzed, tight, fixed to his chest, bent at the elbow, and fisted at the hand. This position protects the paralyzed limb but limits basic functions such as bathing, dressing, and eating. Garrett was placed in casts that progressively straightened out his elbow, an effective but painful process.

"I didn't know anything about brain stem strokes," Eileen said, "and it was a good thing, because I never put a ceiling on his recovery."

As Garrett went through the rehabilitation program at Gaylord Hospital, she learned more and more.

"Garrett didn't even have enough movement to point to a picture board for his basic needs. The first pictures he pointed at expressed hunger and thirst." Eileen looked down, then said, "That was the worst experience for me as a mother: finding out that he was hungry and thirsty." His weight had gone down to 112 pounds. Garrett's rehabilitation physician immediately increased his feeding schedule, and his nurses chased him down in the gym to feed him. He stopped losing weight. Once Kris said it was safe for him to eat soft foods, Eileen brought him chocolate mousse, cheesecake, and other favorites.

In physical therapy, Garrett's first challenge was relearning how to move in bed. Turning toward the right was relatively easy, as he could use his left arm to a limited extent. As for the other side, he squirmed and twisted his trunk and almost threw himself across the bed, but it was such hard work that he had to rely on the staff.

His physical therapist brought Garrett to the gym and placed him on the therapy platform, a wooden plinth with a foam mattress. Awkwardly, fitfully, Garrett learned how to strengthen and control his trunk muscles so that he could sit up without anyone's support. His highly developed balance mechanisms, from years of skating on the ice, helped him improve faster than his therapist expected. But his right foot and knee remained weak and tight. At rest, his foot was pointed down, and his knee was straight; this common pattern of spasticity after a stroke helps the body achieve an upright position.

Once Garrett could sit without any support (almost a month after arriving at Gaylord), he tried to stand. At first, the physical therapist had to lift him up with a waist belt, and he almost fell on her. Gradually, he advanced to standing with minimal help; then he entered the parallel bars with a therapist behind him and another in front. Almost one-legged, he forced himself up with his once-powerful left quadriceps and grasped the bar on his left. Even though his speech was still soft and strangled, he was able to communicate that he wanted to stand on his own, if even for a second. The therapist in front smiled and nodded at the one behind Garrett, and they let go. Garrett stood.

Then his therapist suspended him in a harness above a treadmill so he could relearn how to walk. It was an emotional sight for the Mendez family. Eileen cried. It was inconceivable that Garrett had improved so fast. She had never been so proud of him. Now Garrett knew that he would walk again. He did not know how long it would take or what devices he would need, but he would not be in a wheelchair for the rest of his life.

Even with all the efforts of his physical therapist, Garrett's right foot did not move. In addition to stretching and strengthening exercises, she stimulated his leg muscles with tapping, vibration, and electricity. Recovery in a limb after a stroke typically occurs first in the proximal muscles, and Garrett fit that pattern. When the right foot remained paralyzed, the therapist obtained a plastic brace that supported Garrett's foot as he took his first steps in the parallel bars. Even though his right knee was unstable and almost buckled, he reached the end of the bars. The applause was as rewarding as if he had scored a winning goal. It had taken more than a month to reach this point, and Garrett knew he was still in the first period of his rehabilitation.

By the beginning of February, he walked almost independently with a walker that was modified for his weak right arm; as his brain stem healed, his right leg began to cooperate. Garrett's team encouraged him to use mirror therapy so he could see and adjust his movements with visual feedback. But Garrett refused to look in the mirror, even when brushing his teeth and washing his face. It was hard to see his new self.

On February 9, 2006, eight weeks after arriving at the hospital, Garrett walked out the glass doors. His right leg was still weak, so he needed a four-legged walker and the ankle brace to lift up his foot. Otherwise, he would have stumbled and fallen.

"He walked out of Gaylord on his own," Eileen told me. "It wasn't pretty, but he did it."

On remembering this moment, Garrett grinned. His satisfaction with this victory over paralysis spread across his face to include his carefully trimmed mustache and goatee. I imagined Garrett holding on to his walker mostly with his left hand, his spastic right hand grasping a built-up handle, his right leg dragging even with the help of a plastic brace that lifted up his foot, his body leaning toward the left in order to swing his right lower limb forward, his heart racing with the effort, and his lungs filling with the air of independence.

When he left the rehabilitation hospital, Garrett ate only soft foods. (For almost a year, he had to be supervised at meals because of coughing or choking.) He had to feed himself with his left hand because he had almost no control of his right upper extremity; he couldn't even use it for bathing or dressing. After hours of practice with his speech therapist and many indecipherable letters and words, Garrett learned to write with his left hand. His speech was slurred, for his palate and oral muscles were weak. Fortunately, his vision was essentially normal. He was fully aware of his losses, but he wanted to make the most of his remaining movement and abilities.

Eileen Mendez had expected him to stay at Gaylord for at least six months, so she was thrilled with his progress. But she knew that his journey was far from over.

"They were amazing, absolutely amazing, with what they did for him at Gaylord," she told me. "Then it was up to us to keep his rehab going."

Garrett knew it would be a long time before he got back on the ice. "I wanted a full recovery," he said. "As an athlete, I knew I had the will and the self-discipline to get better."

For Garrett's safety, the Mendez family installed grab bars for the shower stall, a shower chair, a side railing for his bed, and a baby monitor. To improve his balance, Garrett used the Wii at home. The first version of the video-game console, which came out in 2006, had a wireless remote controller and detected movement in three dimensions. Garrett soon mastered the game's tasks; then he tried the 2008 version to further challenge his balance. At this stage of his recovery, he had advanced from the plastic brace to a less restrictive and lightweight carbon-fiber device that fit into his shoe, lifted up his foot, and prevented tripping. Training with the Wii eventually helped him regain a more natural and safe pattern of walking, even with the brace that supported his right ankle.

"I didn't realize it at the time," Garrett said as a smile flashed across his handsome face, "but the Wii also improved my golf swing."

"We often use it at my rehab center," I said. "In fact, the tennis game helped me recover from a shoulder injury."

A recent review of devices like the Wii found improvements in arm function and activities of daily living, such as bathing and dressing.[3] Larger studies and longer-term monitoring are needed to definitively confirm the value of virtual-reality therapies for stroke rehabilitation. At present, especially in light of recoveries like Garrett's, this approach seems promising.

Garrett had learned to skate at the Wonderland of Ice, which planned a fund-raiser for him during a tournament on Labor Day weekend. In the final game, Garrett hoped to walk out and drop the puck. His high school coach, John Watson, and his former teammates put his equipment on him and opened the gate to the rink. He barely stood up, and his ankles twisted as his team pushed him around—but he was back on the ice. In the spirit of the fund-raiser, he used the gift from his hockey friends to buy gym equipment for his basement: a treadmill, an elliptical trainer, and five weight-training stations.

"He worked out every day after therapy," Eileen said. "He was so focused and kept pushing himself."

As his balance improved, Garrett graduated from the carbon-fiber brace to the WalkAide, a device similar to the NESS L300 that had helped Pat Wines. When his right heel left the ground and his knee was bent in preparation to swing the leg forward, the WalkAide sensed his intention and stimulated the muscles to lift up the foot and prevent it from dragging on the ground.

Years later, Garrett's gait pattern still had some typical features of a stroke survivor: his overall balance was slightly off, and he made an effort to lift up his foot, which would then come down with an audible slap. With the WalkAide, these gait abnormalities were corrected to some extent (as shown in a YouTube video of him).[4]

Eileen told me that Garrett's speech was limited for two years after the stroke. A turning point came when Rich Coppola, a sports reporter from a local TV station, featured Garrett at a hockey fund-raiser for Gaylord Hospital. Although his speech was not clear, he seemed to enjoy the interview. It was as if a new facet of his personality had emerged.

Garrett knew that telling his story could help others, which gave him confidence.

When I met Garrett, his palate still did not move normally. As a result, instead of all the air going through his mouth when he spoke, some leaked out of his nose. This affected sounds that need air pressure, such as "b" and "p." Garrett was fitted for a palatal lift retainer, a device like a splint, to compensate for this problem. He could wear the retainer in the back of his throat because he had also lost his gag reflex (a rare example of how a neurological deficit can actually be beneficial). Since getting the device, Garrett felt that his speech was much clearer. During our conversation, he had only minor problems with certain sounds. His voice was slightly soft and nasal, but I fully understood him.

His parents and Jenn continued to devote themselves to his rehabilitation. Eileen and Gary were real-estate appraisers who owned their own small business. After Garrett's stroke, they had dropped everything to support and encourage his recovery. Even after their exhausting experience during Garrett's stay in the ICU and his inpatient rehabilitation, they took him to therapy sessions five times each week for the first three years. His family was determined to find the best possible therapy, even if it meant traveling all around the state. For instance, they drove about two hours each way to the University of Connecticut at Farmington for specialized speech therapy and his palatal lift retainer. Following these intense three years of therapy, the Mendez family provided Garrett's therapy at home and researched the newest techniques and technologies for stroke rehabilitation.

"We went through all our savings for Garrett's recovery, and our business suffered," Eileen told me. "We had to give up our real estate work to help Garrett, and then the housing bubble burst. It's not been easy. But getting Garrett back has been worth it." At an especially discouraging time, Eileen joked with Gary and said, "Why can't we just win the lottery?" He replied that, with Garrett and Jenn, they already had. "You are so right," Eileen said.

Garrett's right upper extremity was the slowest to recover: he was able to grasp objects but not release them. A custom-fitted SaeboFlex made his hand more flexible and functional. This dynamic splint, unlike a rigid plastic one, counteracted the tendency of Garrett's hand to tighten up into a fist. The spring-loaded action of the SaeboFlex helped open up his hand after he picked up or grasped objects such as a glass or tooth-brush. When Garrett's hand relaxed, it remained in the best position of function, with the wrist and fingers extended and ready for use. The dynamic splint is based on the principles of neural plasticity. Through numerous repetitions during essential tasks such as feeding and dress-ing, the brain recovers by forming new nerve connections and improv-ing existent nerve pathways.[5]

For the tightness of his hand, Garrett received Botox injections every six months for two years. Then he received them only once a year for two more.

"They helped me keep my hand open," he said. "But they were pain-ful." Garrett was glad when his occupational therapist felt there was no significant benefit and the treatment was stopped.

Even with the SaeboFlex and the Botox injections, Garrett's self-care was limited due to weakness and tightness at the right elbow. A year after his stroke, his occupational therapist, Stacy Melillo, attended a conference in Massachusetts and saw the Myomo (a condensed form of "my own motion"). Melillo realized that the device (see figure 14.1) might help Gar-rett produce more fluid movements of his arm, so she suggested enrolling in a clinical trial. Garrett did not meet the criteria for the study, but he and his family met with Kailas Narendran, who had codeveloped the Myomo.

As graduate students in a 2002 robotics class at the Massachusetts Institute of Technology, Narendran and John McBean set out to design a device to assist any remaining elbow function in people with strokes. Their device was approved by the FDA in April 2007, after five years of research and development.

"The patient has to think about moving and initiate the move," Nar-endran said in an interview.[6] "So the patient is always in control."

Like other advanced devices for rehabilitation after brain damage, the Myomo is based on neuroplasticity: hundreds and eventually thousands of movements would renew nerve pathways and create new connections for brain recovery. By initiating the movement of partially paralyzed arms, the device assists repetitions of muscle action that help rewire the brain. With the Myomo, people relearn how to move their affected mus-cles and restore functional movements of their arms.

"Ultimately the goal with stroke patients is to make it so you don't need devices," Narendran said.

Even two years after his stroke, Garrett had no functional use of his right arm, so the entire Mendez family was amazed when the device helped Garrett raise his arm to play tic-tac-toe on a wall board.

Eileen said, "The look on his face when he could move his arm for the first time in two years was priceless."

Narendran must have realized that he was talking to no ordinary stroke survivor and allowed Garrett to try the device in early 2008— before it was approved by the FDA—as an outpatient at Gaylord Hospital.

Testing with many patients has shown that the battery-powered, computerized brace, which weighs less than two pounds, is effective as early as two days and as late as twenty-one years after a stroke.[7] Surface electrodes are strapped onto the arm to sense the electrical activity of the biceps and triceps muscles, which bend and straighten out the elbow. Software continuously monitors the electrical activity of the muscles. When the patient tries to move his or her elbow, the computer system senses the muscle signal as the intention to move, and a motor assists with the desired movement. This process is so fast that it seems natural to the patient, whose muscles act in concert with the brace. When the patient wishes to stop moving, the device senses that intention and stops its motor.

This process reeducates the biceps and triceps muscles as patients move their elbows in a natural manner, synchronous with whatever muscle function is possible. Through numerous repetitions, damaged brains are retrained and healed. Patients are actively involved in the therapy, which helps create new neural pathways to move their affected arms. Patients with elbow tightness that limits their movements may also find that their range of motion is improved.

Garrett's therapist fitted the device to his arm with padding and a strap to ensure proper placement of the sensor electrode over the muscle group. For instance, to retrain feeding, the sensor would be placed over his biceps. To straighten the elbow and push up from sitting to standing, the sensor would be placed over his triceps. Whirring softly, the robotic device helped him straighten his elbow smoothly and efficiently. The therapist pushed a button, and the device automatically sensed the electrical muscle activity from Garrett's muscle; this procedure also helped the therapist set the level of assistance required to move the elbow, based on his remaining strength. As Garrett used the device and became stronger, the therapist reduced the force provided by the motor in the brace. In addition, the therapist gave him guidance and cues with her hands to ensure that Garrett would develop normal patterns of movement. Throughout, Garrett initiated the movements while the Myomo sensed and assisted his intention to move. It also reinforced his success as he struggled to move his arm, eventually leading to greater independence.

Garrett and I discussed his experience with the electronic brace. Once it was approved by the FDA, he used it on a regular basis at Gaylord Hospital's outpatient rehabilitation program.

"It was a little heavy when I worked above my head," Garrett said. "But my arm was very weak. The therapist held up my arm so I could focus on specific muscles."

"How did it feel when you first used the Myomo?"

"It felt great to control my arm again. I hadn't been able to do that for two years," Garrett said.

"What were you able to do that you couldn't before the device?"

"I straightened my elbow to reach for things, and I bent my elbow to pick them up. My arm used to move in a jerky motion, but the Myomo made it more fluid."

"So you were able to use your biceps and triceps?"

"Yes, my biceps were always firing, but the Myomo calmed down my muscles so I could finish a task."

"It helped your spasticity?"

"Yes, it helped a lot," Garrett said. "I can pick up a laundry basket or a plate with both my hands. I can dress myself more easily. I can use both my hands to pull up my pants. I don't worry about dropping things, such as glasses. I can pour juice more easily."

"So it's made a difference," I said. "Sounds like it got you through that phase when your elbow had weakness and spasticity."

"Yes," Garrett said. "But I don't use the Myomo anymore because my therapist thinks my arm's doing fine without it."

"Great," I said. "There are all kinds of adaptive equipment for disabilities, but the ultimate success of a device is when it's no longer needed."

Garrett's right arm now had functional use and good range of motion, in contrast to when it was almost fully paralyzed. He had almost full elbow extension, and the stroke was apparent only when he used his right hand for fine motor tasks. Garrett was so successful with the device that he and the Myomo were featured in a CBS News segment.

With skilled therapy and the SaeboFlex dynamic splint, Garrett recovered some hand function. He did not require the NESS H200, as in the case of Kathy Spencer. His major problems were elbow weakness and tightness, which limited activities such as eating and dressing. Garrett's only option, short of functional electrical stimulation with surgically implanted electrodes—as with Jennifer French—was the robotic elbow brace.

Myomo's latest version, the mPower 1000 Neurorobotic System, is FDA approved for use in the home or clinic. In designing the new device, the researchers invited Garrett to try the new mPower 1000 in a clinical trial at home. Like the earlier device, its electrodes sense muscle twitches as a patient tries to move; then it helps the movement with a motor in the brace. The new Myomo assists with daily activities and exercises for range of motion and strength, which helps the rewiring process in the brain. The mPower 1000 has computerized controls for easy use and built-in Bluetooth capability for communication with external applications and systems.

In collaboration with a research group at the University of Cincinnati, Myomo has developed automated programs (to push, reach, and lift) that integrate the mPower 1000 into therapeutic activities. Progressively more challenging tasks help the patient regain independence with bathing, dressing, eating, and toileting. Another innovation of the new device is compatibility with myGames, a virtual reality system that encourages patients to enjoy their exercises as they play with both real and virtual objects. Patients can "have fun while performing repetitive movements that have been clinically proven to promote motor recovery," according to Dr. Michael McNeill of the University of Ulster, who helped develop the system. As a result, patients are more intensely engaged in their therapy.[8]

The mPower 1000 also records data on the range of motion at the elbow, total movements, muscle exertion, and the duration of usage. These data can be graphed over time to show progress, which may improve the patient's motivation and compliance with therapy. Therapists have Internet access to these data in order to track and document a patient's evolution.

Steve Kelly, the CEO of Myomo, commented that the new device "is the first step toward our long-term vision to integrate different technologies and treatments that get people moving again, give them hope, and let them be where they want to be—at home." The new electronic brace weighs less than two pounds and costs about five thousand dollars, so the corporation hopes it will be a practical option for patients with neurological conditions.[9] (Health insurance typically covers 80 percent of the cost of devices that are clearly beneficial.)

Gaylord Hospital now regularly uses the Myomo, thanks to the Mendez family and other donors who support rehabilitation for patients with strokes.

"I wanted the Myomo for my son, and I knew it would help other stroke survivors," Eileen said. She wished it had been FDA approved when Garrett was at the rehabilitation hospital. "I think we would have seen a faster return."

Garrett's left side was fully back to normal, his speech was getting clearer, and he didn't need a walker with the lightweight ankle brace. His favorite device for his weak right foot was a tightly laced hockey skate, which acted like a brace as he glided around on the ice—without any signs of a stroke. His physical and occupational therapy continued, twice a week, for he and his family believed that stroke rehabilitation was a lifelong process. For three years after starting the Myomo, Garrett had been driving a modified car; it had a left-foot pedal and a spinner knob on the steering wheel for sharp corners and parking.

"I was a little nervous at first," Garrett said. "I still don't use my right hand to drive. I'm afraid it'll get stuck on the wheel."

"So your hand has some spasticity?" I said.

"Yes, but I control it pretty well."

"How's your walking? Do you still have a limp?" I said.

"Not really, not like before. If I pay attention to my walking and go through what I have to do in my mind, I can walk very well."

"And you're able to skate now. How often do you do that?"

"Once a week for about an hour," Garrett said. "I use a hockey stick to work on shooting and passing."

"How long have you been skating?"

"For about two years," Garrett said.

He said that his mother would not let him play competitive hockey. As a parent, I could certainly understand Eileen's position, so I nodded but didn't comment. His friends from high school and an assistant coach worked with him on the ice.

"Hockey's a lot more fun than regular therapy."

"Sounds like a great way to get rehab," I said.

In addition to his physical rehabilitation, Garrett wanted to resume his education. He worked at a dry cleaner's on a part-time basis, but he left the job to attend Housatonic Community College in Bridgeport, Connecticut, where he took courses such as "Introduction to Computer Applications" for programs like Microsoft Access and Excel. He typed with his left hand and also forced himself to use his right. Recently, Garrett achieved a vocational goal by becoming a physical therapy aide. This job involved cleaning therapeutic equipment, applying heat and ice packs for patients' pain management, and helping patients regain their strength. Garrett also resumed playing golf, which helped him transfer his weight from leg to leg and improved his balance.

"I can play a few holes, but my endurance is limited. I get tired walking long distances."

"How about other hobbies and activities?" I asked.

"Sometimes I go to bars with friends. But I don't do that often—it's hard for me to speak in noisy bars."

Now I noticed that the volume of Garrett's voice had lowered, probably due to fatigue.

"Do you need help with your daily activities?" I asked.

"Yes, I need some help with cutting food and reaching for things on high shelves."

"Garrett has shifted some activities to his left hand," Eileen said. "We never pushed him to eat with his right hand. As a mother, I didn't want him to get stressed over dinner; we enjoy eating together as a family."

I wondered about his daily life, and I asked Garrett, "What's your typical day like?"

"I get up, do my therapy exercises. Then I take a break, hang out for a while. Later on, I do some homework. And I work out very day."

"What do you do?"

"A lot of core exercises, like the bird dog. I also do sit-ups and reach for a ball between my feet. I work on strengthening my right knee, hip, and ankle."

"That's pretty rigorous," I said. The bird dog is an excellent way to strengthen core muscles such as the abdominals and lower back. From a resting position on all fours, a person points one arm, as if toward a bird, then raises the opposite leg.

"Yes, and I also like cardio, an hour every day."

"You're in great shape. What's your height and weight?"

"I'm five-eleven and 165 pounds," Garrett said. "I'm happy with that."

We discussed the reasons for his successful rehabilitation after a severe stroke.

"Hockey has given me the discipline and the motivation to recover," he said. His teammate and best friend, Jeff Velleca, said that Garrett was superb at both offense and defense. He would slide on the ice to block a slap shot and also swing his stick to send the puck flying into the net. Although he admired his friend's hockey skills, Jeff was even more impressed with his character. He recalled that Garrett could stand up only with the help of his teammates when he first skated onto the ice in 2008.[10]

Once a week, Garrett skated as part of his therapy, and he told me, "At first it was really hard, but it's getting easier and easier" (see figure 14.2). His only comment about his losses was, "I used to skate without thinking." Then he gave a wry smile and shrugged off his momentary lapse into the past. Recently, without a walker, Garrett and Jeff skated together, passing the puck back and forth, just as in the old days.

Garrett was not bitter about his life being shattered in his youth. He had shared his experience with large crowds, including students at Sacred Heart University. Within a year of his stroke, he helped coach a high school hockey team. To recognize a former star, who was even more impressive off the ice as a stroke survivor, Notre Dame High School retired his jersey. Garrett was never interested in public speaking and certainly not in talking to large crowds, but he began educating people about stroke prevention and treatment. While speaking to large audiences, Garrett discovered that he enjoyed public speaking and hoped to make a career of it.

Garrett's speech was slightly slow, and he had trouble with certain sounds, such as sibilants and some consonants. But his story was so compelling that I made sure to catch every word. I imagined that all audiences, large or small, would have a similar response.

Rehabilitation would be a never-ending process for Garrett, but he had strong supporters. His family had always helped him, and his parents actively searched out new therapies and devices.

"Even if it helps him one percent, that's great," Eileen said. "We'll give it a shot."

"I'm like you," I said. "I'm willing to try anything that might help, as long as there aren't any adverse effects. There are many paths to recovery."

Garrett also credited his friends with helping him recover.

"They've been with me since day one. But they never help me unless I ask."

"Friends make a big difference in recovering from a stroke," I said.

Garrett's hockey team remained active in his rehabilitation and supported Garrett's mission to educate the public about stroke prevention and treatment. (A number of his videos are on YouTube.)[11] Garrett had participated in the American Heart/Stroke Association's Walk every year after his stroke, along with his family, friends, Gaylord Hospital staff, and hockey teammates. In 2010, Eileen and Garrett discussed their experiences at Connecticut's stroke conference, with an audience of 250 people that applauded their inspiring story.

"He never looks back," Eileen said. "He never says, 'Poor me.' He just thanks God that he got through, and he works every day on his rehab."

In 2010 Garrett was honored to be a Stroke Ambassador for a Heart /Stroke Association fund-raiser in New Haven, and his team helped organize a skating event with local high school hockey programs. When Garrett glided onto the ice, his right leg weakness barely noticeable, his balance carefully controlled (for he had fallen on the ice while relearning how to skate), and a grin visible through his face mask, the audience erupted with applause as if he had just scored the winning goal for the Stanley Cup.[12]

The Bionic Man

Matt, 2005

Matt smiled out from the January 2005 issue of the *Brown Alumni Magazine*, handsome in a high school photograph. Emily Gold Boutilier described how Matt "guided a cursor across a computer screen by merely wishing it into motion." I told the reporter that he was an ideal patient for the study: the ventilator tube blocked a voice-activated system, so the BrainGate had therapeutic potential. In addition, he fit my intangible criteria: a pioneering spirit and altruism. He "wants to give people hope," Ellen said. "This gives him a sense of purpose."

Matt had to testify at the stabber's trial in late January. Any travel outside New England Sinai was a logistical challenge and required careful planning. To be in court on time, he got up at 5:30 AM for his bowel care and to bathe, groom, dress, and eat. Before leaving the haven of New England Sinai, Matt required deep suctioning with a plastic tube to clear his lungs. A urinary tract infection threatened him with possibly fatal sepsis, so Matt was on intravenous antibiotics to prevent this complication of spinal cord injuries. Strapped into his electric wheelchair, which was clamped down to the floor of the Nagles' van, Matt went back and forth to court for about two weeks.

Because of the ongoing trial, I went to see him in the evening. From Providence, I drove north on Interstate 95, past patches of snowcapped evergreens huddled among bare, icicled trees. For the one-hour drive, I listened to Ravi Shankar playing his sitar, the notes swirling around as the musician improvised complex variations of his raga, bending the notes and coaxing the deepest possible resonances. I wondered how this virtuoso musician's brain would appear on a functional MRI scan as millions of neurons fired in harmony and electrical impulses reverberated through his brain.

Near the entrance to C Wing was a painting of three horses rampant on a green field, their snorts almost audible and their hot breath almost visible. A bright sitting room opposite the secretary's desk looked out onto the cedar hedges, which were weighed down by the snowfall. In a small conference room, I sat down to review Matt's latest reports and was pleased to see that all was well. As usual, the corridor leading to Matt's room was lined with oxygen tanks, yellow isolation carts, portable computers for quick access to medical records, and a defibrillator.

After greeting Matt, I checked the skin around the implant to ensure that it was not getting damaged, as it had with the monkeys. Then we talked about the trial. It depressed Matt. As a matter of form, the defense attorney requested a dismissal, because no one had actually seen his client stab Matt. When the judge denied this request, the assailant started crying.

"What was *he* crying about?" Matt said. "He didn't give a shit about me."

Matt gave me some details of the trial. Initially, bail had been set at $250,000, but the judge doubled it. An attorney said that a judge had released this man for a knife assault that caused minor injuries—a week before he stabbed Matt on the beach. Matt and I shook our heads. This career criminal could have been in prison on July 3, 2001, if that judge had been less lenient.

"It's the worst injury I've seen from a knife," said Matt's trauma surgeon, Dr. Hirsch, during his testimony.

Matt found out that, during a heated discussion, some jurors had started shouting. To reach a verdict, they asked the judge to clarify "assault with the intent to kill."

"Well, he didn't mean to only scratch you," I said.

"My family couldn't look at that knife," Matt said. "I almost got sick to my stomach when I saw that blade."

Almost four years after the stabbing, the knife still gave Matt nightmares. I imagined the courtroom gasp when the knife was displayed, its blade gleaming with violence, the missing knifepoint still embedded in Matt's neck.

"He's looking at ten to twenty years, in a place with murderers and rapists," Matt said. "Hard time."

"You're doing hard time, too," I said softly.

"Yeah, a life sentence." Matt moved his head up and down, the only gesture he could make for emphasis. Except for the ventilator's puffing as it inflated and deflated Matt's chest, the room was silent. I placed my hand on Matt's shoulder and changed the subject to his research accomplishments.

"The implant looks fine, no drainage. And Abe said the recordings are great. You've become a pro with the BrainGate."

"We learned as we went along," Matt said. "A special time in my life. I'll never forget it—never."

"We want to try a few other things," I said. "Dr. Donoghue has some ideas about improving the software. To make it more efficient for you to use the BrainGate."

"I'm willing to give it a shot."

Exactly a week after my visit, on February 8, 2005, the stabber was found guilty of armed assault with intent to kill and sentenced to ten years in prison with ten years of probation—the maximum sentence. Dennis Tatz of the *Patriot Ledger* wrote that the "incident dramatically changed his life and the lives of his family members but had not defeated [Matt]."

At trial, Matt testified, "You got me that night, but I tell you, you won't beat me. I'm not going to live my life as a loser like you." This comment drew tears from the stabber. Then Matt added, "I can't believe I am sitting here in this chair. I look out the window and say, 'This is my life.'"

In her testimony, Ellen Nagle said, "You took away his capacity to have a career. You took away his chance to have children. . . . You took away his chance to walk down the aisle with his bride. You took away his ability to breathe without a machine . . . sit without being strapped in a chair, sleep without being turned every four hours, eat without being fed, or even stay alive without ingesting more than sixty pills a day."

Matt's condition and involvement in the BrainGate study were regularly reassessed. In early February 2005, we held the quarterly meeting of the clinical oversight committee. It included David Apple from Shepherd Rehabilitation Center and Steve Williams from Boston University, as well as Gerhard, Burke, and Shawn. We reviewed Matt's clinical data over the past seven months and unanimously agreed to continue the study. Due to the initial lack of nerve recordings, Cyberkinetics developed an impedance tester for the BrainGate system. The tester sent a small current into the patient to check the electrical continuity of the microelectrode array and its bundle of wires. This potentially risky procedure required approval by the FDA and New England IRB as well as a new informed consent form.

In mid-February, I gave a presentation at the annual conference of the Association of Academic Physiatrists. Leaving the snowfall behind, my family and I flew off to Tucson, Arizona. The first night got down to sixty degrees, and Nita and Jacob dragged me out to the outdoor hot tub. During breaks in the conference, we swam, played tennis, and practiced putting on the golf course. The presentation to my colleagues elicited a spectrum of responses, from amazement to disbelief. I wished that Matt had been there with the BrainGate for a live demonstration. It would have been a wonderful break for him, but his medical condition made air travel risky, almost impossible.

Shortly after I returned from Arizona, the FDA and New England IRB approved the impedance tester and new informed consent form. Shawn wanted to use the tester as soon as possible, so the next day I explained to Matt how the device would check the BrainGate circuit with a small current that passed through him. There were minimal risks, but I could not guarantee complete safety.

I had wanted the Nagles there for the informed consent process, but Matt said it would be an extra trip for them, so a friend signed on his behalf. Clearly impatient with the process because he had already read the form, Matt made occasional comments, such as, "That's fine, Doc." Still, I went through the entire process, as required by the FDA and our IRBs. (This is yet another example of the difference in perspective between patients and physicians. Desperate for any form of recovery, people with severe disabilities are vulnerable to the next "miracle cure.")

While I examined Matt, he asked me to adjust the tilt of his wheelchair, give him a drink, check the heating system (because he was warm), and scratch his scalp around the pedestal. With gloved hands, I took care of all these requests, uncertain of the next time Matt would have someone at his disposal.

When finished with all these requests, I asked, "Will that be all, sir?"

Matt smiled and said, "Thanks, Doc."

<hr />

Later in April, at the annual meeting of the Rehabilitation Association of New England, I asked the audience to cross their arms and legs. Then, while they remained in that locked position, I asked them to pick up a pen from the table.

"No cheating," I said. Some people laughed at the impossible request, and an enterprising person tried to bend down and pick up a pen with his teeth.

"That's how it feels *all* the time for someone with tetraplegia," I said. "And that's why I started working with the BrainGate project and Matt Nagle."

The conference goers were interested in other research to help people with spinal cord injuries. In the spring of 2005, stem cell research blossomed in a profusion of publications. Researchers from Wisconsin and Sweden reported that injections of neural stem cells into rats with spinal cord injuries decreased their pain and increased their movement. Scientists at the University of Wisconsin–Madison used embryonic neural stem cells to stop the progression of amyotrophic lateral sclerosis (ALS, or Lou Gehrig's disease) in the spinal cords of rats. Another team from Wisconsin used specific growth factors and human embryonic stem cells to create functional motor neurons that could be "replacement motoneurons" for spinal injuries.

<hr />

During a visit in early May of 2005, I learned from a nurse at New England Sinai that Matt had another urinary infection, a severe one that required two intravenous antibiotics. Fatigued and barely able to open his eyes, he hadn't slept for a long time. His legs were difficult to move because he had more spasticity than usual, a result of his inflamed urinary tract. Matt was in considerable pain, but he helped me do the examination required for the protocol and our quarterly FDA report. We talked briefly about the study, which Cyberkinetics wanted to continue with Matt. It was unclear what would happen in June, exactly one year from the time of his implant.

As summer approached, it was time either to remove the implant or extend the study. Cyberkinetics wanted to continue collecting data and to improve the software. So I sent an e-mail to New England IRB with updated information. Inclusion criteria were that patients should *not* have "any significant health changes that would increase the risk of skin infection," and, most important for Matt, *not* have "another implanted stimulator." He still wanted a pacemaker for his lungs so that he would no longer be tethered to the ventilator; that would be impossible if he kept the BrainGate implant. As before, the form listed all the possible adverse events.

New England IRB recommended minor changes to our consent form, and the FDA said that "the patient must have usable recordings during the last three months" of the study. The FDA also wanted us to inform patients about alternative therapies such as "sip and puff or head switches, speech recognition systems (Dragon systems), EZ Keys/Plus, and Eyegaze tracking systems." Once we satisfied these requirements, both the FDA and New England IRB approved the extension study.

This was a good time to step back for a panoramic view and answer our initial questions. The BrainGate implant had successfully recorded electrical signals from Matt's brain. There were no safety issues. Matt had controlled the cursor by merely thinking about it—in a consistent and reproducible manner. He had also used the BrainGate to control a television, play video games, open simulated e-mail, open and close a prosthetic hand, and control a robotic arm.

With the extension study, our team hoped to improve the software and to enhance Matt's control of the BrainGate. But a problem arose: Matt's diet, lack of activity, and weight gain had led to diabetes, which increased the risk of a skin infection. A way around this criterion would be my authority, as the study's clinical investigator, to make the final decision. In contrast to the first informed consent form, the new one for the extension study was only eight pages long.

After reviewing the pros and cons with Matt, I said to him, "You've already made a historic contribution to people with disabilities. I'd like us to do more experiments and improve the software. But I have an ethical responsibility to inform you of the higher risk of a brain infection— because of the diabetes."

Matt nodded and said nothing.

"Fewer than half the microelectrodes are giving signals, and we don't know exactly why. And you can't get the phrenic nerve pacemaker while you've got the BrainGate. I know you want to get off the vent."

I was silent, and the vent, as if on cue, hissed into Matt's lungs. We looked at each other and smiled. The ventilator had spoken: remove the BrainGate.

After much conflict, I had said what was best for Matt. Cyberkinetics would lose a patient with a functional implant, and I would lose consulting fees and the chance for further groundbreaking research. But as a physician in the fullest sense of the word, I could make only one recommendation to this young man, with whom I had formed an intense researcher-patient relationship.

Matt was silent for a minute. Then he said, "I hear you."

For a while, we chatted about our summer plans, and I promised to be in touch soon for his final decision. Matt's answer was clear, however, and I was relieved after leaving. I had done the right thing for Matt, and that was all that mattered. On the way home, I called Burke to give a quick update. I respected Burke's opinion and was pleased that he, too, did not wish to pressure Matt to extend the BrainGate study.

The summer of 2005 was full of activity related to stem cells, in both the scientific and political realms, and Matt watched the events unfold as he would a football game. On May 20, President Bush vowed to veto any legislation that eased current limits on stem cell research. Four days later, the House of Representatives passed a bill to lift some restrictions on federal funding for stem cell research, with the support of fifty Republicans. Within two days, the corresponding bill was introduced in the Senate and backed by senior Republicans. As promised, President Bush vetoed federal funding for research with in vitro fertilization (IVF) embryos that were destined for destruction. At first, Matt was angry, but his hope for stem cells was not extinguished.

"I can wait," he said, "until Bush is out of the White House."

In contrast to the president, some states were more supportive. Connecticut allocated $100 million for stem cell research over ten years to help its biotechnology companies compete with California, New Jersey, and Illinois.

Matt and thousands of people with spinal cord injuries also found hope in the research of Dr. Hans Keirstead at the Reeve-Irvine Research Center in California. First, he transformed stem cells into oligodendrocytes. These cells form myelin, the insulation that speeds up nerve impulses. Then Keirstead injected the cells into rats seven days after incomplete spinal cord injuries. The oligodendrocytes matured, migrated to the damaged neurons, and wrapped myelin around them. Most important, these rats improved their walking, compared with injured rats that received no cells or were treated at ten months. Geron Corporation provided the stem cells and planned the first human study of embryonic stem cells for spinal cord injury in 2007.

Before my next visit, in June 2005, I stopped off at Dunkin' Donuts for Matt's favorite coffee: vanilla with milk and artificial sweetener. He grinned and winked as I walked in with the cup, taking me back to our first meeting. In that moment, I knew he would discontinue the Brain-Gate study and pursue other avenues for improving his independence.

"I know what you're going to say," I said as I sat down.

"Yeah, I want the surgery to take it out." Matt was hooked up to his ventilator and gasped out his answer in a hoarse voice.

With this matter settled, we watched the Red Sox game. They weren't doing very well, but who could complain after their World Series championship the previous year?

"Schilling's having a rough day," Matt said. "Bases are loaded, no outs."

"Oh, that's no good," I said. "What's the count?"

"Three and one. He's struggling; he's about to walk."

"Schilling's a hero—no matter what he does," I said. "He could walk five in a row. Everyone would forgive him because he won us the World Series."

"Yeah, the ultimate Red Sox," Matt said. "That sock with his blood will be in the Hall of Fame someday."

"It sure belongs there."

Seeing Matt upright in his electric chair made me worry about his skin getting worse, and I remembered Christopher Reeve. Sitting for long periods in the chair during the trial had left Matt with a slow-healing pressure ulcer.

"Do you adjust your position periodically?" I asked him. "Tilt yourself with the sip-and-puff device?"

"Every half hour for about a minute."

"Good."

"I've got to watch my skin. I have to be my own best advocate. That's very important if you're quadriplegic."

"You're right," I said, and I looked up at the TV. "Nice curve on that one."

"Imagine doing that when you're almost forty," Matt said. "Oh, come on, just one more like that."

"Three and two, and the bases are loaded."

"Oh, boy. Got to go with the fastball. Oh, shit!"

"Can I quote you, Matt?"

"Sure!" Matt laughed. "The pitch was in there. He had to go for it."

"The Orioles just scored. It's all over."

"It's a shame. Bases still loaded, only one out. Doc, can I have some more coffee?"

"Sure. I would've gotten you a large one, but it might throw off your bladder. How's that going, by the way?"

"I get catheterized twice a day, morning and night," Matt said.

"Only twice a day? How're the volumes?"

"I get about eight hundred cc's. I also void on my own, so I wear a condom catheter and a leg bag."

"I'd rather you were cathed three times a day. To keep the volumes lower."

"It works. Doc, can you scratch my right eyebrow?"

"Sure, how's that?"

"Good. Thanks."

For a while we talked about Matt's life before and after the stabbing, the metamorphosis into a tetraplegic patient in a long-term care facility, the life that was shattered, and the new life that he was still trying to build.

I glanced up at the TV. "Hey, the Red Sox took the lead while we were chatting."

"We're up four to one?"

"Yes, we are."

A panel of statistics flashed across the screen, and Matt quickly grasped what had happened.

"Manny hit a double, and David Ortiz also got a double."

"Those two keep us going, don't they?"

"Oh, yes, definitely. I look forward to the games, you know? It breaks up Saturday pretty good for me. That's tough sometimes."

Shawn informed me that Gerhard wanted to meet with Matt for a pre-operative evaluation and to review the surgical informed consent form. About two weeks later, I was surprised to hear from Tim that Burke was leaving Cyberkinetics for another biomedical firm. He praised Burke as a leader in the BrainGate study and reassured me that Cyberkinetics was still committed to the research. The very next day, Nora Wood at 360usainc.com, an "online source for the disability community," asked if enrollment was still open. I offered to discuss the study with her website members.

Min-Young Paik, researcher at the Korean Broadcasting System, wrote in August for an interview "to film [the] brain chip reading Matthew Nagle's mind and sending [his] thoughts to a computer." She needed my help with getting access to Matt, so I contacted the Nagles. I also told them about my advice to discontinue the study.

"He told me," Ellen said.

"The longer the implant's in his brain, the greater the risk of infection," I said. "The connector isn't fully sealed off. A skin infection could travel down to the brain. And he has diabetes. Besides, he wants the phrenic nerve stimulator to breathe on his own."

"Yes, Matt told us about your discussion," Pat said. "He said he trusted you."

It turned out that the Nagles were leaving soon for vacation and would be away until August 15, so we planned the explant surgery to occur after that date. Matt had done so well that the entire team agreed it would be best to complete the study and avoid any future complications. As living proof of the efficacy and safety of the BrainGate device, Matt was a prize "exhibit" for patients, scientists, investors, the FDA, and the world.

Matt returned to Rhode Island Hospital on October 18, 2005, for removal of the BrainGate. He still had MRSA in the nostrils and armpits (and probably would be colonized with this resistant organism for the rest of his life). We used Bactroban to suppress the bacteria. All the preoperative labs were fine, except for a diabetic glucose of 170. Cloudy urine with many red and white blood cells suggested an infection, and Matt was given the antibiotic Zyvox.

Next morning, with Matt submerged under anesthesia, Gerhard made a ten-centimeter incision, retracted the skin on Matt's scalp, exposed his skull, and dissected out the wire bundle. Next, Gerhard removed the titanium plates and screws that had fixed the wire bundle and repaired the craniotomy site. Bone had regrown at the craniotomy margins, so Gerhard cut it away with a high-speed drill.

Then Gerhard moved the bone flap to expose and lift away the hard dura and the weblike arachnoid membranes, revealing the four-millimeter-by-four-millimeter array still implanted in the motor cortex. Gerhard removed it, noting that the "brain tissue appeared completely healthy, with no signs of scarring, excessive bleeding, or infection." At one area, bone had grown over the wire bundle, so he again used the high-speed drill. Then Gerhard dissected the bundle out of the soft tissues and removed the titanium screws. He pulled out the pedestal, wire bundle, implant, and reference wires as one unit. It was stored in a sterile container for the biomedical engineers at Cyberkinetics, who were anxious to analyze the device and to explain the decline in the number of recordable neurons.

Gerhard irrigated the surgical field with antiseptic and cauterized some small, leaking blood vessels. He then repaired the dura membrane with Neurolon sutures and placed a DuraGen patch for watertight closure. With a titanium plate and screws, he fixed the bone flap back in place. Finally, Gerhard irrigated the wound, stopped all the bleeding, and closed the wound with absorbable sutures for the scalp and nylon sutures for the skin. Matt had no complications.

After the explant, Matt was alert and still able to shrug his shoulders, a movement controlled by cranial nerves and not his spinal cord. He seemed comfortable with intravenous Dilaudid for pain and Ativan for anxiety. His white scalp dressings did not have a protruding implant, unlike the first time I had seen him in the ICU.

His eyes slightly glazed from the anesthesia and sedation, Matt said, "Is it out, Doc?"

"Yes, you're doing fine. Your vitals are OK, and the dressing looks good."

"Thanks, Doc."

Matt closed his eyes, drifting back into postsurgical semiconsciousness. I patted his right shoulder and went to talk to his parents.

Later that day, Matt's voice was dulled, and his nurse noted that he was using more pain medication than expected. He had developed tolerance to narcotics. With Fioricet, Matt's headaches became less intense. Gerhard was not too concerned, but we agreed to watch Matt closely and to keep each other informed of any changes. That night, I went to bed wondering if Matt would have to go back to the operating room. There had been no change in his neurological status, so I felt somewhat reassured. But in someone with complete paralysis of his limbs and no sensory function below the shoulders, there would be few signals of a surgical problem. Fortunately, Matt's pain decreased, and his mental status improved the next day. He returned to New England Sinai Hospital, and as usual, the nurses there did a great job of managing his care and keeping me apprised.

In a month, Matt's wound had healed completely, and hair loss was the only remnant of the BrainGate implant. He looked forward to the day when his dark brown hair could be brushed over the circular reminder of the craniotomy. Matt took a well-deserved rest from everything other than his daily routine. The BrainGate media hubbub had subsided. Now, Matt could resume his search for whatever pleasures remained in merely existing as a fully paralyzed young man.

With Matt's BrainGate journey officially over, it was time to publish our findings. Leigh Hochberg, a neurologist at Massachusetts General Hospital and a postdoctoral fellow in John Donoghue's lab, was an ideal leader for this demanding project. Starting in November 2005, he sent the research team a number of versions of a paper titled "Neuronal Ensemble Control of Prosthetic Devices by a Human with Tetraplegia." Due to its innovative nature, the BrainGate project was given an expedited review by *Nature*.

Typically journals obtain critical evaluations of research articles from at least three objective, outstanding scientists. All of the *Nature* reviewers gave helpful responses, but they also had some concerns. One referee wanted to ensure that we did not "overstate data" and asked for mathematical details. Another did not like the description of Matt as being able to draw a "circle" because his "erratic attempts finally managed a crude approximation. This, of course, is quite a remarkable achievement in itself." So Leigh described it instead as "an approximately circular figure."

Our final article described data from fifty-seven consecutive recording sessions over nine months. Matt's nerve signals ranged from "well isolated single neurons to mixtures of a few different wave forms." On average, the electrical signals were seventy-six microvolts (a millionth of a volt), and twenty-seven nerves gave recordings each day. But the number of neurons could drop down to three, which would not allow Matt to use the BrainGate system.

We described Matt's many accomplishments, which culminated in successfully using a multijointed robotic limb to grasp and transport an object. With all these tasks, Matt was a quick study. He performed them even during conversations and without continuously monitoring the computer screen, important features for people with severe disabilities. For skeptical readers, we included the videos of Matt with the BrainGate.

Near the end of the article, we acknowledged that the system tethers the person "to a bulky cart and requires operation by a trained technician. A wireless, implantable, and miniaturized system combined with automation will be required for practical use. Emerging and available technologies appear to be sufficient to overcome these obstacles, although the challenges of creating a fully implantable system may be formidable."

In conclusion, we optimistically stated that the electrical "activity from neuronal ensembles can provide a control signal after spinal cord injury sufficient to perform at least basic operations for a human with tetraplegia, justifying further engineering efforts."

When the paper was published on July 13, 2006, Matt appeared on the cover of *Nature* with his faithful clinical engineer, Abe Caplan, and the headline "Turning Thoughts into Actions."[1]

After the explant surgery, I visited Matt and performed the final detailed examination required by the protocol. There was no change in his paralysis, sensory loss, spasticity, hyperactive reflexes, and bowel and bladder incontinence. When I finished my notes in the last case report form for Matt, we chatted about his social life.

"You know," Matt said, "I'm not going to lie. It's tough being in this situation. I had so many friends in the beginning, and then they tapered off. Now only four people see me once or twice a week."

For a while, only the ventilator spoke.

"Yes, it's tough," I said. "You were a good friend, and you expected more from people."

"The ones that stuck around are great. I wouldn't trade them for the world."

"Who are they? I know I met one of them."

"Danny Taylor was a grade ahead. We used to go to the high school games, and we became friends. You know, it's funny. We weren't even best friends, but he's stuck with me for more than four years. He comes up with dinners and feeds me. He's got a great heart."

"He's a devoted friend," I said. "There are very few of those around."

"And that's true even if you're not disabled."

"As far as friends go, you once told me about a girlfriend. . . ."

"She was my girlfriend for a year. She's gone. That's all I can say."

"Yes, I understand. Disability changes everything."

"To truly understand what it means to have a disability," Matt said, "you have to be in that person's situation. To explain it is difficult. I thank God for family, friends, and doctors and nurses who care. Being in this position takes a toll on me. I've got to be mentally strong day in and day out. I've got to be constantly aware of my surroundings. I worry about skin sores; I wake up at two and six in the morning to be repositioned.

It's a terrible situation. A bad person did a bad thing. I have to live with it for the rest of my life. But I don't plan on being in a wheelchair forever. I'm hoping that within three years I'll be out walking around, and that's what I believe."

"You mean with stem cells?"

"Stem cells or anything else that comes around," Matt said. "You never know."

⸻

When I visited Matt one evening at New England Sinai, his nurse worried that he wasn't getting enough pressure relief while using his wheelchair. His condom catheter occasionally came off, causing his urine to further damage his skin. On Matt's back I saw a reddened, open sore, and Christopher Reeve flashed through my mind.

Matt lay on an air mattress to minimize skin pressure. The Boston Celtics were playing, and I asked if we could turn off the volume. But we could not silence the ventilator as it forced air into Matt's lungs. His toes pointed away from him, a sign of increased tone and tightness in his legs, and he had a pillow between his knees for his overactive thigh muscles. I asked if his spasticity had worsened.

"Yes, Doc, they had to increase the baclofen." Matt sighed.

"May I check your range of motion, or will that bother you a lot?" I asked.

"No, go ahead. See what happens."

As I bent Matt's right foot up toward him, it started beating rhythmically as if he were tapping his foot. This abnormal response also occurred with his left foot. Matt's knee only bent to ninety degrees. When I placed his pale and paralyzed legs in their usual position, his right knee bent, almost as if he had voluntarily moved the leg.

"Did you see that? *I* did that, Doc." Matt grinned at me.

It was only a hyperactive reflex, so I smiled back and shook my head. Only Matt Nagle would try to pull a trick like that on a rehabilitation doctor.

"The spasticity is bad today," I said. "It's probably the skin breakdown." His brain could not register the pain of a pressure ulcer, but the tissue damage sent shock waves through his nervous system and caused his muscles to shake in spasms. "Are you getting enough pressure relief?" I worried about his skin, and my expression was accusatory.

"Sure, Doc." Matt frowned for emphasis.

His nurse wanted me to persuade Matt to use an indwelling bladder catheter, to avoid urine leakage that would damage his skin. But once again, Matt did things his way.

As a formality, I checked Matt's sensation; I doubted it had changed. He had sensory function at the upper shoulders, and I asked if he felt my hand.

"Yes, but I wish I didn't feel a damn thing there. The pain in my neck and shoulders gets real strong. There's only so much Oxycontin I can take."

I was saddened that this young man, left with so little neurologic function, was willing to give up his ability to feel. Matt's shoulder pain was both neuropathic and musculoskeletal. Neuropathic pain typically occurs in the zone where sensory function has been lost, and patients describe it as burning or tingling. Perhaps the body compensates for the utter loss of sensation by increasing the sensitivity to pleasurable as well as painful sensations. Musculoskeletal shoulder pain results from the paralyzed arm falling out of its socket joint, causing small tears of ligaments and tendons and creating a vicious cycle of pain. Furthermore, spasticity and soft tissue contractures cause the shoulder joints to tighten up. Patients with chronic shoulder pain require troughs on their wheelchairs to support their arms. When moved around, they are vulnerable to the well-intentioned efforts of people who pull on their arms in misguided attempts to help.

Patients with spinal cord injuries have many physical reasons for severe pain, but, in my experience, their psychic pain is worse. Matt and I discussed his pain management: high-dose Oxycontin, other narcotics, baclofen for spasticity, and an antidepressant.

When the phone rang, I held it near Matt's ear. Excited about being at the Celtics game, a friend asked if Matt could see him on the TV.

"Yeah, I see you—upper right corner," Matt said quite convincingly, before laughing at his gullible friend.

While they talked, I looked around the small room. It was decorated with calendars of partially nude women. For a young, virile man, losing sexual function was a tragedy. As far as I knew, Matt did not engage in any sexual activity, but whatever his physical function, he clearly was still interested in the opposite sex. Before Matt needed nitropaste for dysreflexia, to reduce his blood pressure, Dr. Bloom had agreed to prescribe Viagra. He never found out if it worked for Matt. I never asked.

After the phone call, I asked Matt about his signed photo with Britney Spears, taken during a backstage visit.

"I had the vent in me, so I couldn't talk to her," Matt said.

"So you couldn't flirt with her."

Matt smiled, shaking his head with regret.

⸻

Meaghan Murphy was among the friends who visited Matt at New England Sinai, but she hadn't seen him for almost a year because she was away at school. Over the phone he described his success with the BrainGate, but he also admitted that his medical condition was going downhill. As she had done before, Meaghan visited and fed him a favorite meal: Buffalo wings drenched in blue cheese dressing. While Matt licked his lips, Meaghan realized that food was his greatest source of sensory pleasure. She also saw that he had gained a lot of weight; even his handsome face was puffy. But his endearing smile, his laugh, and the twinkle in his eye were bright as ever. Somehow, he had preserved the will to survive and fight. There was still some hope.

Later, she realized what none of his family or friends could say: he was not going to live forever. Matt had been treated for urinary infections, pressure ulcers, and diabetes. Meaghan felt angry, then sad. Matt had been one of the most handsome guys she knew. And now he looked like this.

From Christopher Reeve's book *Still Me*, Meaghan learned about the work, the stress, and the cost involved in caring for a person with tetraplegia. She wished that she had seen Matt more often, even if just to shave his face or feed him.

⸻

Because of Matt's paralysis, medications, medical complications, and depression, he slept poorly. Matt still couldn't fall asleep, so he stayed awake at night and slept during the day. Sleep was a chronic struggle, but now it was worse than usual. Dr. Bloom adjusted Matt's medications and reduced his Valium to help his sleep, but that worsened the spasticity. Managing more than thirty medications for Matt was a delicate and unpredictable balancing act; I admired Dr. Bloom's skill.

Matt's sleep cycle was also disrupted every five seconds by the ventilator. He continued to dream about breathing on his own, free of the ventilator that inflated his lungs with breaths at a controlled frequency and volume of air. In addition to the ventilator in his room, he had a portable one mounted on his wheelchair. Two plastic hoses connected his throat to the ventilator, a noisy and incessant reminder of his disability and dependence. The hoses were just as strong as the paralysis that bound him to his wheelchair, and Matt yearned to cut them off.

Months after his injury, Matt had undergone surgery to check the phrenic nerves in his neck, their most accessible location. These two nerves, one on either side, are controlled by the spinal cord at the upper cervical levels, the area where Matt had been stabbed. Doctors had thought perhaps a pacemaker could be placed there to stimulate his paralyzed diaphragm and help him breathe on his own. But the surgeon said, "There was so much scarring where he was stabbed, I couldn't even find the nerve. We had to give up."

An alternative was to implant electrodes to stimulate the phrenic nerves near the diaphragm, far away from his scarred neck. Even while Matt had the BrainGate, we had discussed this surgery, which required cutting through his chest wall. The procedure was risky, I advised him, and another implant was not allowed by the protocol. Now Matt wanted the pacemaker so he could breathe the New England air on his own.

The phrenic pacemaker is a small box that sends radiofrequency signals through a wire antenna to an implanted receiver. In turn, this device sends electrical impulses to electrodes at the phrenic nerve, causing the diaphragm to move at the typical respiratory rate of twelve times per minute. Similar to normal breathing, the lungs are rhythmically aerated, but lung volumes are low because the muscles of the rib cage are paralyzed.

Matt returned to Rhode Island Hospital on November 7, 2005. His neurosurgeon, John Duncan, cut through Matt's chest wall on both sides to place the electrodes that stimulated the phrenic nerves. The operation went smoothly, but he had severe pain and a fever for two days. To everyone's relief, a chest X-ray was normal. His temperature came down, and his pain lessened. He was sent back to New England Sinai Hospital.

Thrilled to be free of the ventilator's plastic tubes and incessant hissing, Matt told me, "The pacemaker stimulates my diaphragm. I can breathe with it twenty-four seven. It's given me a lot of freedom. But once in a while I get sick, and they have to put me back on the vent."

Off the ventilator, Matt now had less air moving through his lungs, so his speech was hard to understand. His voice remained soft, and he still clicked his tongue to get someone's attention at a distance of even a few yards. Scarred by the ventilator's tube in his throat, Matt's vocal cords were weak and damaged. Part of his problem was coordinating his speech with the pacemaker's cycle. Ellen had heard Christopher Reeve talk with a similar device and noticed that he breathed in concert with the pacemaker. But Matt just didn't have the patience, and he talked right over it.

Ellen e-mailed me to say that Matt was "having great luck with the pacer—off for thirty-six hours at a time." After all this time on the ventilator, Matt's regaining such an elemental function was remarkable. She also mentioned "a great piece on *60 Minutes* last week on Dr. Hans Keirstead and stem cell research. If it weren't for scientists, there wouldn't be any hope. And for Matt and others like him, there is no life without hope."

⸺⸺⸺

In April 2006, I received an e-mail from a patient who had been tetraplegic since 2002. His sensory function began to return almost three years after his injury:

> As the months pass, the tingling gets more intense. My doctors . . . moved my right foot, and I told them I could feel the tingling shift as they moved my foot! Is it at all possible for me to volunteer for any of your studies or rehabilitation exercises? Anything would be greatly appreciated!

E-mails such as these, full of energy and enthusiasm, filled me with hope. If, after three years of paralysis, this young man believed in the power of science, certainly we clinicians and scientists should be inspired by this optimism.

For the Nagles, the phrenic pacemaker made a big difference when they brought Matt home on weekends. But, as with any medical device, it had to be closely monitored. Matt occasionally woke up feeling short of breath; if this happened, he used a call light activated by turning his head to call his parents. Once, he couldn't breathe in spite of the pacemaker, so Ellen and Pat had to speed him back to the ventilator at Sinai, bagging him all the way to force oxygen into his lungs.

Research on phrenic nerve pacemakers was published more than forty years ago, which shows how long it can take for basic science to become clinically useful. Only in 1994 did a study prove that pacemakers were superior to ventilators when it came to electric wheelchair mobility and communication.

Many Americans became aware of prosthetic implants through a 1970s television show called *The Six Million Dollar Man*. A severely injured and barely alive astronaut was introduced by a voiceover in the opening credits: "We can rebuild him; we have the technology. We have the capability to make the world's first bionic man. Steve Austin will be that man. Better than he was before. Better. Stronger. Faster." Reconstructed, the man became a secret agent who used the superhuman strength and speed of his bionic limbs and the hyperacute vision of his artificial eye to save America from all enemies, including aliens and deranged scientists. Even before the phrenic pacemaker, there was the dream of replacing vital structures, repairing the body, and creating a "better" human. The fictional Steve Austin's bionic right arm, both legs, and his left eye cost $6 million.

At its 2006 conference on experimental biology, the American Association of Anatomists announced its "vision for a bionic man, where biology is merged with electronics." Possible advances included "a robotic arm that can play the piano, a bionic eye . . . and a powerful 'external skeleton' that confers superhuman strength." A superhuman exoskeleton was not just science fiction: Berkeley Bionics was founded in 2005 to develop the research of the Robotics and Human Engineering Lab at the University of California—for soldiers to carry two-hundred-pound loads over rugged terrain and for people with paralysis.

Research scientists may have visions and dreams, but clinicians have to struggle with the numerous complications that affect people with severe disabilities. For more than three years, Dr. Brian Bloom watched over Matt's medical status. They had a wary yet friendly relationship, one that sometimes flared into arguments over Matt's medications or fast food or cigarettes and at other times erupted into shared laughter over a joke. Of all the complex and challenging patients that Dr. Bloom had treated, Matt easily took the prize. When Matt became diabetic— and insisted on guzzling high-calorie Mountain Dew—Dr. Bloom struggled to keep him free of diabetic complications that could have led to vomiting, hypotension, and death. Matt's pain medications were also a frequent source of friction.

"Sometimes I gave in about the narcotics," Dr. Bloom told me after the BrainGate study. "There was some drug-seeking behavior, but he had real pain, and I felt for him."

Two physician assistants had helped care for Matt, but one left because of conflicts over narcotics for pain. When Matt wanted more opiates, a nurse said it had to be cleared with Dr. Bloom. Enraged, Matt tried to attack the nurse with his electric chair.

"He scraped my leg. It was painful." The nurse then added, "He couldn't take a swing at anyone."

Once Matt received the phrenic pacer, Dr. Bloom worried about his airways. The device made him more independent but didn't aerate his lungs as well as the ventilator.

"I can't remember how many times I treated Matt for bronchitis," he said on one of my visits. Furthermore, because spinal cord injuries damage bladder function, Matt continued to have frequent urinary tract infections. Repeated antibiotic treatments breed resistant bacteria, and Dr. Bloom said, "He was colonized with some bad bugs."

Paula Picard, the nurse manager on C Wing, also got to know Matt well over the years, and I chatted with her one day about his care. Worried about his fast-food diet, she tried to reason with Matt, but he responded, "Talk to the head," because he couldn't raise his hand.

His schedule, his meals, his medications: everything was a negotiation. Picard had learned to handle a lot from patients, but if he swore, she left and returned in ten minutes. Many things, big or small, would upset Matt, but Picard had treated enough people with severe disabilities to understand their psychodynamics, their frustrations over having no control of even basic bodily functions. So she and the staff tried to comfort him. The nurse manager and I had a long chat after the phrenic pacemaker was placed.

"He went through a lot," Picard said. Tears came to her eyes. "Matt was very upset when he couldn't get the pacemaker at first because of scarring at his neck. Luckily it worked when Dr. Duncan went through the chest. But he didn't help himself. Matt had the best wheelchair and the best cushion, yet he got a bad pressure ulcer. There was a lot of time in the chair—with no pressure relief. Some people with quadriplegia can adjust, like our long-term patients. We had a woman who lived twenty-five years on a vent. I think women are less into control, and they care better for their skin."

"Young men have the hardest adjustment," I said.

"They can be tough. Once I told Matt he couldn't do anything illegal. It looked like he was using extra Valium, so we held his dose." The nurse paused and dabbed her reddened eyes. "He pushed the limit with everything: hockey, football, the BrainGate, meds. . . ."

Matt also pushed for stem cell therapy in South Korea and asked Katie Perette to go with him.

"Someone's actually walking over there after stem cells," he said, referring to Hwang Mi-Soon. "What do I have to lose? I want to try everything."

She tried to reason with him. "But Matt, that's only one patient. Maybe you should wait, see how others do."

"Katie, I'm stuck in this chair—this damn cage. What do I have to lose?"

"Matt, how do you know the stem cells are safe?"

"I don't, but someone has to be the crash-test dummy. I'm not saying no to anything."

"I give you a lot of credit," Katie laughed. "You never stop trying."

"I'm going to walk. There's no doubt in my mind."

But when Katie visited on other days, despair had replaced his determination.

"Katie, pull the plug on me. *Pull the plug on me!*"

"No, Matt."

"I hate you."

"Matt, there's a reason you're here. You're not giving up. That would be a cowardly move. You have to fight and fight."

He settled down and, later in the visit, said, "I want to go to the Miami Project."

At this renowned rehabilitation center in Miami, patients received exercise with stimulating electrodes that kept their paralyzed muscles moving in a coordinated manner, with both physical and psychological benefits. Katie was relieved to hear that Matt not only stayed abreast of the latest spinal cord research but also had a specific plan; it would give him the strength to continue fighting.

Without the ventilator tube in his throat, Matt could now use his voice-activated system to surf the Internet, write e-mails, and enjoy the music on MTV.com. But even voice control was laborious, because his voice was soft and raspy with the phrenic-nerve pacemaker. Although Matt gave Dr. Bloom the impression that he was quite pleased with the device, one day he confided to Katie that he was embarrassed about his voice and hated its sound. Still, Katie was happy for Matt's success.

Patience was never a virtue for Matt, but eventually—after some frustration—he learned to coordinate his voice with the electric shocks to his diaphragm. He still clicked to get Katie's attention at times, like when they were watching a ball game with the volume turned up.

Katie tried to keep him involved in the outside world. With time, she had fewer episodes of crying in her car, shaded by the pink-blossomed crabapple trees in the parking lot at New England Sinai.

|||||||||||||||||||||||||||||||||||||||

I learned from Danny Taylor that the prosthetic arm had been a favorite BrainGate activity for Matt, but its use was limited by the protocol. Matt had lived for movement—on the football field, at tailgate parties, on the beach, or in bed—and now he constantly dreamed of moving by any means. He hoped that the BrainGate would be able to move his own arm after he controlled the robotic one. In fact, Matt told Danny that we could amputate his arm and attach a robotic one, which would then be controlled by the BrainGate. He wanted to become the bionic man of the new millennium.

"He joked about it, but he was dead serious," Danny said. "He'd do it. He was a physical guy, and when he lost that, he lost everything."

"How's he been doing since the pacemaker?" I asked.

"That made a huge difference. It gives him freedom."

Not having a ventilator also allowed Matt's nurse and Danny to occasionally take him to Foxwoods Casino in the wheelchair van. Matt loved blackjack and parked himself at a low, wheelchair-accessible table; Danny sat by his side to play his hand.

"Once he won about two hundred dollars, but he must have spent a thousand to do that," Danny said. "He loved to gamble but never had much luck."

Casino facilities weren't always accessible for Matt, but he made his way. Once he got close to a blackjack table using the sip-and-puff mechanism on his wheelchair. Then he asked an Asian man to slide over so he could get closer. Perhaps there was a language barrier, or maybe the man was just ignoring him.

"Excuse me, please slide over," Matt repeated.

When there was no response, Matt backed up his wheelchair and rammed his way in. He played blackjack that night.

Imagining a God

Matt, 2006

In January 2006, *Wired* ranked the BrainGate among "the top five scientific and technical discoveries of 2005."[1] The magazine described Matt as the "first neuro-cybernaut," the "first paralyzed person to control an artificial hand by brain power alone."

Proud of his pioneering role in the BrainGate study, Matt contacted Dr. Chae, the South Korean researcher who had treated Hwang Mi-Soon with stem cells. Paralyzed from the waist down, she walked more than ten meters with crutches after his therapy.

Matt's e-mail had the subject heading "Paralyzed man looking for help—desperately!" He wrote: "They tell me it may take seven years to get FDA approval for the BrainGate to move my arms, to move my legs. I can't wait that long, and I know that stem cells are going to be the healing agent I need to help me walk and use my arms again." Matt asked Dr. Chae to Google his name and wrote: "I beg of you to give me an operation that could save my life. I have read your miracle story, and I would sign my life away if you would consider me for the surgery. Please contact me as soon as possible."

Dr. Chae's assistant sent information about Histostem and his stem cell treatments. In the e-mail was a table with headings for various diseases, name, age, sex, efficacy, and side effects. Of three patients with spinal cord injuries, one recovered enough "to distinguish hot and cold," "to move her left fingers, and [to] defecate by herself"; none had any adverse effects.

Detailed technical information lent credibility to the Histostem project. Dr. Chae claimed to have treated people with "spinal cord injuries, liver cirrhosis, diabetes, ALS," and many other diseases. The e-mail mentioned forthcoming publications "in well-known journals," from which Matt could learn about "how we operated [on] the patient Hwang Mi-Soon and her medical conditions and how she progressed." Dr. Chae's assistant wrote that "the results have been successful for all."

To enroll as a patient, Matt had to send a blood sample to match his white and red blood cells to umbilical-cord stem cells; that would reduce the risk of immune-system rejection. In addition, the doctor requested Matt's medical history and CT or MRI scans of his spine. Informed consent would be obtained "for Korea FDA's approval." Finally, Matt would be notified of the treatment date, location, and the logistics of traveling to Korea. A disclaimer stated, "Histostem does not warrant the success of the treatment. Unfortunately, a cost to cover the medical and laboratory expense will incur, and it will be estimated and advised before any treatment."

Matt asked for my opinion. Any rehabilitation physician would have been thrilled with these miraculous recoveries—but I wondered about their scientific validity. Even if the stem cells stimulated regrowth, how would the nerves properly reconnect across the segment severed by the knife? Thousands of nerve fibers had to reconnect for Matt to regain any function. How much scarring had occurred where the knife had slashed through Matt's spine? And the knife point embedded in Matt's neck would have to be removed. I tried visualizing the stem cells entering the point of injury, stimulating nerve regrowth across the dead tissue, healing the wound, and ultimately producing electrical impulses that would move Matt's arms and legs. My imagination failed me.

Matt and I chatted by phone. "I'm being purely scientific," I said. "Let's forget about politics and religion. These are only case reports. They're not verified by independent scientists or published in reputable journals. There may be some theoretical benefit, but I can't recommend this treatment."

I hoped Matt realized that my primary concern was his well-being. To avoid bruising his optimism, I agreed to assist after he carefully considered this project.

Ellen thanked me in an e-mail: "I know Matt can be challenging at times, but he respects you very much, as do we."

She knew I didn't favor this stem cell program, but I offered to e-mail Matt's medical records to Dr. Chae and to arrange a spinal MRI scan. If he passed the preliminary screen, I would get the red and white cell typing for a match with umbilical-cord samples in Korea.

Matt was entering an arena fraught with medical, ethical, and operational issues. Obtaining and sending his medical information to South Korea was simple, but an overseas trip for Matt—with tetraplegia and a phrenic pacemaker—would be a complicated project. Using umbilical-cord stem cells would avoid the ethical issues of embryonic stem cells. But there was no guarantee of any benefit. At this stage, scientists only dreamed of the day when stem cells were effective and safe for treating spinal cord injuries in humans.

Ellen and Pat Nagle reinforced my concerns about stem cell therapy in South Korea. We all asked Matt to be patient. So much exciting stem cell research was being conducted in his own backyard—at Harvard University and by Boston biotechnology companies. I contrasted the scientific and ethical standards of South Korean and American research. In the United States, approvals are required from institutional review boards and the FDA, which oversee clinical research.

"What if there are complications?" I said to Matt. "Will Dr. Chae come here?"

Matt only shook his head. Having been through so much, including the BrainGate study, he knew what I was implying.

After some soul searching, Matt decided not to proceed with Dr. Chae's stem cell treatment. Ellen, Pat, and I breathed a sigh of relief.

South Korea had been a leader in stem cell research. Dr. Hwang Woo-suk at Seoul National University transferred a cell's nucleus into an egg to create human embryonic stem cells, as described in *Science* in 2004. A year later, in the same journal, he triumphantly published the first creation of stem cells derived by merging donor eggs and cells from patients with Parkinson's disease and spinal cord injuries. This major breakthrough would allow autotransplantation, in which the stem cells would contain the patient's own DNA, with minimal risks of rejection by the immune system.

South Korea rewarded Dr. Hwang with lavish laboratories and research funding. To the international scientific community, he was a rock star. Then, an investigation found that the informed consent process for the egg donors was unethical, and the stem cells described in both *Science* articles were actually from frozen embryos. They were not created by transferring the nuclei of cells from patients to donor eggs. Dr. Hwang retracted his articles in disgrace, but he had already raised the hopes of millions of people with diseases and disabilities. Matt became more cautious about stem cells after this scandal erupted.

During one of my visits in the winter of 2006, he said to me, "Thanks for giving me a heads-up on Korea." But he continued to dream of experimental therapies, including embryonic stem cells.

"Stem cells can't come fast enough for me," Matt said. "I'm just waiting for it, praying for it."

"Have you been in touch with anyone else since Dr. Chae?"

"No. But there's a bill in Congress. Maybe I can get treated here. I think the next president's going to be in favor of stem cell research. We'll see what happens."

"Everywhere I talk to people with disabilities, they ask about stem cells."

"What do you think?" Matt asked me.

"Stem cell therapy will happen someday. I don't know if they'll be adult or embryonic or genetically engineered. The studies have to be done correctly—ethically and scientifically."

"It'll happen because there are so many people who want it to happen," Matt said.

"Millions have to live with MS, Parkinson's—and spinal cord injury."

"The public doesn't realize it," Matt said. "Life in a wheelchair is difficult. It's a struggle, one day to the next. It is so hard. . . ."

Silence fell upon us like the cold and heavy clusters of snow that slowly covered the landscape outside.

"Stem cell research really frustrates us," Pat Nagle said. "The jacket for Matt's golf tournament says 'Miracles for Matt' and 'Support Stem Cells.' Some people ask about it, and I try to explain. But others just don't understand. They ask if babies are being killed for this research. Then I tell them about stem cells from the umbilical cord and the nose."

"Some people just hear the politicians and the propaganda," Ellen said. "We're lucky to live in Massachusetts, with Harvard and other institutions that have private funding."

"Oh, yes, it's going to happen," Pat said.

"And all these other countries are working at it," Ellen said. "I mean, there's an international effort. Everybody wants to be the first, you know?"

"Some are doing a better job than others," I said. "England has a large bank of stem cell lines. Sweden and Israel are also doing a lot of research."

"And Matt advocates for himself," Ellen said. "He's out there looking. If Harvard was doing research, Matt would be pushing to go there."

I asked Matt if he had heard about the rats with spinal cord injuries that got stem cells. He hadn't, so I described the April 2006 publication by Margot Mayer-Pröschel and her colleagues at the University of Rochester and Baylor University. Spinal cord injuries cause scars that impair nerve regrowth, leading to bowel and bladder dysfunction, weakness, and sensory deficits. The researchers transformed stem cells into astrocytes, which are the nourishing matrix cells of the central nervous system. The astrocytes repaired the spinal cord injuries by preventing scars and encouraging nerve regrowth. In contrast to rats with placebo therapy, the astrocyte-treated rats had better mobility. More than half of the damaged nerves were regenerated by this therapy and formed a bridge across the injury only eight days later.[2]

"That's great," Matt said, "but what about stem cells for people like me?"

We both knew that stem cell therapy for spinal cord injuries would require thousands of hours of research, millions of dollars, and government support. US policy, developed by President George W. Bush, allowed federal funding only for research with stem cell lines in existence prior to August 2001, a month after Matt was stabbed into paralysis. But the private sector, including universities and corporations, continued studying new human embryonic stem cell lines that might lead to effective therapies.

Perhaps due to his many frustrations, in addition to stem cells, Matt continued his sports gambling and got deeper into debt. His friend Danny told me that he had tried stopping a few times, but it was the only excitement in his life. Always a fan of the sports channels ESPN and NESN, Matt relied on the television for entertainment after he became paralyzed. "Wouldn't it be great if I had four different TVs with all the games on?" he once said to Katie. He knew all the current trades, especially for his beloved Boston teams. But the information didn't help Matt's betting, and he lost more than he won.

When he couldn't pay up and the bookie harassed him, Matt said, "Stop it or I'll get my family on to you. They're cops—state cops. Drop it unless you want to get caught."

Then the bookie threatened Matt with physical violence.

"What are you going to do?" Matt laughed. "Break the legs of a guy stuck in a wheelchair?"

He laughed again and said, "I don't feel a damn thing there anyway."

But the bookie continued his demands.

"OK, are you going to stab me?" Matt said. "That's been done, too."

Angered by these taunts, the bookie called the hospital and threatened Matt's nurses if they didn't get Matt to pay up. Faced with this escalating situation, the facility moved Matt to a less accessible and safer place: the chronic care ward.

Ellen and Pat happened to visit Matt the day after his transfer and found an empty room. A janitor gave them Matt's location, and they rushed to his new room. Matt appeared comfortable, but the walls were bare, and his belongings were in boxes.

"What happened?" Pat said.

"Oh, I wanted to come here." Matt offered no other explanation.

"When did they do this?" Ellen said.

"Middle of the night."

"The middle of the night?"

Now Pat realized what had happened. The disagreements between Matt and the staff must have gotten worse, so they decided to move him. He laughed.

"Did they put a sack over your head and get you out of there?" he asked.

Ellen shook her head. She knew firsthand about Matt's problems with caregivers.

"Next time," Pat said, "you'll be out in the parking lot."

That was the real story behind the midnight ride of Matt Nagle, but Paula Picard politely told me, "We send our longer-term people there if they're stable."

In his new dwelling, Matt continued researching stem cell therapy for spinal cord injuries. He would ask a nurse to set up his voice-activated computer system and start surfing the Internet. The *Boston Globe* reported that Beijing neurosurgeon Dr. Hongyun Huang implanted olfactory cells obtained from aborted fetuses—at a cost of twenty thousand dollars. These cells are like astrocytes, the protective covering of neurons.[3]

Then Dr. Bruce Dobkin and his American colleagues reported the first outcomes study of this treatment in *Neurorehabilitation and Neural Repair*.[4] Having examined the paralysis and sensory loss of these patients before and after surgery, they were unconvinced. Of their seven patients who received the olfactory cells, none recovered. Dr. Huang reportedly accused the US physicians of being liars.

Patients and some caregivers perceived minor improvements, but this could have been due to a placebo effect. Five patients had major complications, including meningitis, pneumonia, and gastric bleeding. Prospective patients seemed oblivious to the risks. Although a patient was able to walk, she had bladder dysfunction, so Dr. Huang drilled a hole in her neck to inject the cells. It is unlikely that the FDA or any institutional review board—I serve on one—in the United States would have approved this risky experimental procedure.

In the summer of 2006, Matt was delighted to hear that former first lady Nancy Reagan, the widow of a conservative president with Alzheimer's disease, supported stem cell research.[5] She said, "There are so many diseases that can be cured or at least helped." Also that year, actor Michael J. Fox said that it "hurt to see the president use the one veto of his administration to strike down this legislation." Conservative radio show host Rush Limbaugh commented on the actor and his struggles with Parkinson's: "He is exaggerating the effects of the disease."[6]

Some of Matt's friends understood his frustration with disability and his desire for a cure through stem cells.

"The way I look at it," Katie Perette told me, "we get up in the morning, have a shower, do whatever we want. And then there's this kid, sitting in a wheelchair, watching the clock tick away, saying to himself, 'Three hours and my friends will be here. We'll watch the Red Sox, just like the old days.' And no one shows."

"Disability changes everything," I said. "I've seen what happens to the social lives of my patients. Very few friends are like you."

"Matt was the best of friends," Katie said, "and not just to me. If he met someone and the person asked for a loan the next day, he'd find the money. He was always very generous. We had some great times together." She paused and smiled. "You should see my pictures of Matt doing kegstands." An image came to mind of Matt suspended upside down in the air by friends who grasped his legs, his muscular arms extended to support himself on the keg, guzzling beer from a hose.

I wondered if Matt's sense of hopelessness affected his self-care. One of his nurses told me that his poor dental hygiene led to sore gums and painful cavities. Narcotics helped, but Matt needed a dentist. So Dr. Bloom found one who treated tetraplegic patients.

On the appointed day, Matt said, "I don't feel like going, Doc."

Canceling the visit would make him jump through more hoops, so Dr. Bloom negotiated with Matt, trading a privilege for keeping the appointment.

"That was the only way to handle things," Dr. Bloom said, shaking his head in frustration as we talked. "He has to have control, even if it's to his own detriment. His diet's bad, so his weight went up. He loves takeout: calzones, pizza, burgers. Boxes of Cap'n Crunch and bottles of Gatorade are lined up on his windowsill; he tells me they're for friends."

"Matt's physical pleasures are limited to his mouth," I said. "There's nothing below his shoulders."

Following the comments of Nancy Reagan, who had a number of Republican supporters, Matt became more optimistic about stem cell therapy. During the summer of 2006, we discussed the ethical issues involved in stem cells for spinal cord injuries. I referred him to an illuminating article by Jeffrey Rosenfeld and Grant Gillett, two neurosurgeons who discussed the ethics and mechanisms of stem cells.[7] First, through chemical and physical signals, they could help axons cross the gap at the injury site. Second, they could develop into neurons that repaired the damaged area. Finally, they could encourage neuronal regrowth. Ideally, labs would collaborate to develop the optimal combination of these therapies.

In addition to the fundamental ethical issues of stem cells, there were moral questions related to repairing the spinal cord. One had to be cautious in applying results from animal experiments to humans. For instance, young animals might have had better results than older ones due to the neuroplasticity of youth. Movement due to reflexes had to be distinguished from functional recovery of movement. Finally, Rosenfeld and Gillett recommended that treatments within two months of injuries should be distinguished from later therapy, since the repair process was quite different for those two groups. (Almost all of the rat studies gave stem cells soon after injuries.)

The neurosurgeons believed that clinical trials with stem cells should include a control group of patients who undergo a sham surgery to address the placebo effect. In fact, decompressive surgery relieves pressure on the spinal cord and should be considered. Patients also had to be informed of two other hazards. The immune system might reject stem cells, but immunosuppressant drugs create the risk of tumors in the distant future. And genetically engineered stem cells might eventually lead to cancer.

Matt understood firsthand how the tragedy of spinal cord injury led to extreme pressure for treatment by patients, families, biotechnology companies, universities, and politicians. I gently reminded him that scientific objectivity could be compromised by hubris and the pursuit of power, money, and fame—as in the case of Hwang Woo-suk in South Korea.

In reporting medical discoveries, researchers and the media had to avoid creating unrealistic expectations in patients. Matt's experience with the media was educational. False hope was an easy pill to swallow, but it carried many side effects. The miraculous potential of stem cells for repairing spinal cord injuries was linked to challenging ethical and experimental issues that had to be carefully studied before treating patients. Clinicians should first protect the well-being and rights of patients.

Ellen and Pat appreciated my advice for Matt as he pursued his quest for recovery. It was an intense process for the entire family, so one day I asked how they were handling the stress. Pat said that they had a time-share in Aruba. It was their favorite vacation spot, a home away from home.

"After Mattie got hurt," Ellen said, "we felt guilty about going."

"Hon, we need to recharge our batteries and stay healthy and happy, or we won't help him at all," Pat said.

"We feel bad, you know, we're away and . . ." Ellen's voice trailed off.

"He can't come," Pat said. "But I make her go every year."

"It's a good idea," I said. "Too many families are destroyed by disabilities like spinal cord injury and MS."

"In the beginning," Pat said, "we felt bad about the littlest things. We didn't like going out for dinner because Mattie couldn't—things like that. You can't do that to yourself. You've got to stay healthy and happy the best you can. Then we can help Matt. And we take him out with us."

"Where do you go?" I tried to imagine Matt's large electric wheelchair inside restaurants.

"His favorite place is the Monponsett Inn, about forty minutes from here," Ellen said. "He loves to sit on the deck overlooking the lake."

"On sunny days, we use their gazebo, put on his sunglasses, watch the ducks, and have a beer," Pat said.

"You know those little things?" Ellen said. "Oh, that makes us so happy."

"He loves baseball, but Fenway's too difficult," Pat said. "So we go to the Paw Sox games in Rhode Island. I love it there. Great park, Triple-A ball, easy access. Before the game, I find out the best place for Matt and the van. I talk to the cops and the box-office people. Then we're ready; I know he's going to get in and get out. We see the game and go out for dinner. But I have to do all that recon work."

"Even after the Americans with Disabilities Act," I said, "it's hard for people in wheelchairs to get around."

Ellen nodded.

"Sometimes we get invited to a handicapped-accessible place," Pat said. "I go there and find just a rail in the bathroom. It's an older building, not under the new guidelines. They don't understand. They plan to pick him up and carry him in. Then I tell them the chair's three hundred pounds and Matt's two hundred. Nobody wants to lift him anymore."

In addition to trips to Monponsett Inn and baseball games, Matt enjoyed being outdoors in the pastoral setting of New England Sinai

Hospital and Rehabilitation Center. In the summer of 2006, parked in his wheelchair in front of the hospital, Matt was immobile. Sunlight streamed onto his face while the pear tree's pink blossoms twirled in the wind, clouds floated across the sky, seagulls flew toward the ocean, ants marched in single file to scavenge a doughnut crumb, a red cardinal flashed into the pine tree, visitors for other patients strolled inside with flowers, a squirrel raced up a crabapple tree, and blades of grass waved in the wind.

That day, Dr. Bloom found Matt with a cigarette miraculously lit in his mouth. He had no way to get help if the cigarette fell on his lap and set him on fire or a spark exploded his oxygen. But he was happy, content. Dr. Bloom didn't ask about his accomplice but explained that this was a bad idea. Matt was able to enjoy most of the cigarette anyway.

<hr/>

Later that summer, a priest offered Matt the sacrament of the Anointing of the Sick from the book of James. Matt was grateful but asked about embryonic stem cells.

"Matt, your life is sacred, and what was done to you was a great evil. But we cannot correct that by doing something else that's wrong. A good end does not justify an evil means. Embryonic stem cell research destroys human life."

Matt was silent. The only audible sound was an IV pump beeping down the hallway. "What about the embryos in IVF clinics? Thousands are being thrown away."

"The Church looks at an embryo as a child of God," the priest said.

"What about aborted fetuses? What about getting stem cells from them?"

"Unintentionally, that might encourage or minimize the effect of an abortion by saying that something good is coming out of it. Matt, we have to think of your spiritual life as well as your physical suffering."

Matt was familiar with the research of Dr. Hans Keirstead, who injected embryonic stem cells into rats with spinal injuries. The cells helped repair the nerves and stimulated their growth in the injured spinal cords. Matt knew that Geron Corporation, which funded this research, planned to perform human studies in 2007.

Matt saw stem cells as his salvation. He said, "If anything can heal me, stem cells can." For him, the Holy Grail was not a golden chalice, ornamented with sacred engravings, but a plastic petri dish that held pure neural stem cells—of any origin—that his body would not reject.

A former altar boy at Saint Francis Xavier Church, Matt had discovered the mysteries of the glittering wine decanter, the embroidered silk raiment, and the fragrant smoke of the silver censer. One day after Mass, Ellen was pleased to hear that eight-year-old Matt felt he had a special relationship with God.

Matt told me that he could imagine a God that created heaven and earth, a God that was almighty, eternal, and beyond measure. Reaching back to his altar boy days, Matt imagined a God that was infinite in intellect, will, and perfection. Matt could even imagine a God that allowed his attacker to destroy his life as a vital and happy young man, a God that allowed him to be cut off from his body below the neck, a God that—as a priest said—used his tragedy to bring many souls closer to heaven and to teach people about the value of human life.

But Matt could not imagine a God that allowed the creation of embryos for infertility, then allowed excess embryos to be destroyed. He could not imagine a God that permitted thousands of embryos to die to make a point about the sacredness of life instead of using them for stem cell research to possibly ease the suffering of countless people. Nor could Matt imagine a God that denied medical treatments with stem cells created by nuclear transfer. If his own DNA were merged with eggs donated by a woman—perhaps Katie—those cells might heal his spinal cord.

Matt tried to imagine a God in whom he could believe.

A Starship Trooper's Electronic Suit

Floyd Morrow

On a warm day in August 2006, Floyd Morrow was working hard as a mechanic supervisor near Atlantic City, New Jersey. He was glad when his company's store manager invited everyone to take the afternoon off for a trip on the corporate yacht. They sailed on the serene waters, enjoying the warm day and an air show. Three miles offshore, one of the men lost his footing on the fly bridge and fell seven feet—directly onto Floyd. When it happened, he was partially bent over on the deck, and the force of almost two hundred pounds crushed his lower thoracic spine. He was instantly paralyzed and lost all sensation from the waist down.

Floyd was taken by the Coast Guard to the Atlantic City Medical Center, where surgeons cut open his back and placed metal rods to stabilize his fractured spine. In doing this, they hoped to preserve his damaged nerves. After the surgery, Floyd was monitored in the intensive care unit for ten days.

At first, he couldn't believe what had happened to him and asked himself, What did I do to deserve this? His legs were still paralyzed. In the intensive care unit, Floyd had a lot of time for soul searching. He concluded, If I have to live like this, I'll make the best of it. God has a plan for me.

Once it appeared that he was stable and would not succumb to complications of the spinal cord injury—such as bleeding, infections, and blood clots in the lungs—Floyd was sent to the Bacharach Institute for Rehabilitation in Pomona, New Jersey. While at this center for more than two weeks, Floyd learned to adapt to paraplegia and the associated problems of a spinal cord injury, including bowel and bladder dysfunction. His legs had no movement, and Floyd had to relearn how to move around in bed. Almost six feet tall and 195 pounds, he had played football, basketball, and track in high school, so he had excellent upper body strength. Now, he needed a wooden board on which to slide his paralyzed lower body from his bed to his wheelchair and back. By the end of his stay, however, Floyd could do this with his own arms.

Although his legs had no voluntary movement, they moved on their own due to spasticity. Floyd's lower motor neurons, going from the spinal cord to the muscles, were no longer controlled by the upper motor neurons that went from his brain's motor cortex to his spinal cord. Fortunately, these random movements were not severe, and Floyd did not need medications or surgical procedures. Some patients with severe spasms require high doses of medications or even a pump that delivers one of these drugs (baclofen) directly to the spinal cord.

One of the biggest problems related to paralysis and sensory loss is skin breakdown. Unable to move their bodies, patients develop pressure areas; because the sensation of pain is absent, they are unaware of problems. Once pressure ulcers occur, patients have to limit their mobility and keep their weight off the affected areas. If open ulcers become infected, they can lead to fatal infections. So Floyd and his wife, Maryann, a nurse at a nearby hospital, were careful to watch out for pressure areas.

After coming home from the rehabilitation center, Floyd was in his wheelchair with his legs strapped to the leg rests to prevent them from falling off and getting injured. In spite of his precautions, the straps caused blisters that Floyd didn't even feel because of his absent sensation. When the blisters opened up, they formed ulcers. In the average person, these areas often heal on their own, as long as they are kept clean and do not become infected. But paralysis often leads to swollen legs and poor circulation, which compromises wound healing.

As a result, Floyd had to be treated with a wound-vacuum system.[1] In theory, negative pressure promotes wound healing through a variety of mechanisms: it removes inflammatory and infectious drainage, reduces swelling, increases blood flow, and promotes regrowth of healthy tissue. First, the skin is cleansed, which may require surgical removal of dead or infected tissue. Then, a foam patch is applied to the open area, which is sealed with a semiporous protective layer. Finally, a circular patch that is attached to a suction tube is placed on the area. The tube is connected to a machine that applies a low, constant vacuum.

After Floyd received this treatment, however, skin grafts were necessary; and his mobility was often compromised by these procedures.

In March 2011, Floyd and I discussed how life had changed for him, his wife, and their two daughters since that summer day in 2006. Even in a wheelchair, he appeared like an athlete: slim but well muscled. Many patients become heavier after losing the ability to walk, but Floyd said that he had lost weight; unfortunately, it was mostly muscle. He insisted on staying independent and active in his wheelchair, which was hard work and required a lot of calories. The Morrows had to replace a step with a ramp so that he could wheel inside the home. Their bedroom was moved to the back of the house. To create entrances for Floyd's wheelchair, they also had to remove the doors from the hinges. For these renovations and copayments for medical expenses—which included medications and rehabilitation equipment—the Morrows had no financial help.

Floyd talked to his company's president, who agreed to buy him a van. The health insurance company paid for modifications such as hand-held controls for braking and accelerating. Floyd got the van a year after his injury, which improved his independence, but the nature of his job was such that he could not return to work. Before getting the van, he transferred in and out of his Jeep, and it was difficult for his wife to fold and stow the wheelchair.

Floyd later got a titanium wheelchair that weighed less than twenty pounds. Titanium is three times stronger than aluminum, and his wheelchair was designed for heavy use and uneven surfaces. But it could not be folded, so traveling was a problem. Floyd was an active rider, unafraid of rough terrain, and his adventurous rides could be risky. In fact, Floyd had fallen out of his chair three times. Twice he flipped over the chair, in the process tearing his right shoulder's rotator cuff. As a result, he had to have surgery to reattach the supraspinatus muscle.

Mobility, at any cost, was essential for Floyd. As a star athlete in high school, he loved the exhilaration of sports. As he grew older, Floyd stayed in shape through his active job, carrying parts for boats and toolboxes, climbing ladders, and walking constantly. Experienced in the construction industry, Floyd also took care of home repairs and renovations. In fact, the summer before the accident, he had redone the wooden flooring and the interior trim for the Morrows' house.

Floyd yearned to walk once more, if even a few steps. In December 2008, Floyd's physical therapists gave him a pair of regular leg braces that traveled from his waist down to his feet, immobilizing and supporting his knee and ankle joints. Fabricated from metal and plastic, these braces are cumbersome and static; their role in walking depends entirely on the paralyzed patient, who relies on his or her arms and rotation of the trunk muscles. A study of patients with paraplegia who used these braces found that their energy consumption, compared with people without a disability, is eight times greater. The cardiac workload becomes excessive, even with short distances. The study concluded that walking with braces is "not a valid alternative to wheelchair propulsion."[2]

But Floyd's upper body strength and determination were reasons for optimism, so he started using the braces in March 2009. Once the knee mechanism was locked, Floyd was able to stand and walk in the parallel bars for an hour each session, three times a week. With the coaching and encouragement of his physical therapist, Floyd used the braces for more than four months. On occasion, he even used them at home, without the help of any therapists. But Maryann worried he would fall when she was not there, so he stopped using them.

"We got out of the habit of putting them on," he told me.

Floyd gave up his dreams of walking again until his doctor received a letter about the ReWalk, a battery-powered exoskeleton that might help him walk. She asked Floyd if he was interested, and he said, "Yeah, it sounds like fun."

Exoskeletons are usually associated with insects, but in Robert Heinlein's 1959 novel *Starship Troopers*, he imagined a futuristic armored suit that greatly enhanced the trooper's strength, speed, and endurance. Heinlein's "powered suit" also conferred superhuman abilities with vision, hearing, and communication.[3]

A little more than a quarter century later, a US Army Ranger named Monty Reed fractured his spine in a parachute accident. Partially paralyzed, he read Heinlein's novel while in his hospital bed. The book inspired Reed to design an exoskeleton called the Lifesuit, which he eventually built in 2001, fifteen years after his accident. (That year he also created his nonprofit research organization, They Shall Walk.) Successive models improved upon his prototype; the twelfth version (2005) was good enough that he used the electronic braces in the Saint Patrick's Day Dash in Seattle. At the beginning of 2012, he was working on the sixteenth version of his Lifesuit.[4]

In 1997, almost forty years after Heinlein's book was published, an Israeli engineer and entrepreneur named Amit Goffer became completely paralyzed after a car accident. With doctorates in electrical and computer engineering, Goffer designed and built exoskeleton trousers for people with paraplegia. He founded Argo Medical Technologies to develop and market the ReWalk. Its rigid, motorized components support the lower extremities, helping patients with paralysis at the hips, knees, and ankles to stand up and walk with crutches. Sophisticated motion sensors are connected to a computer in a backpack worn by the patient. As the patient leans forward, the sensors detect the movement and "realize" that it is time to start walking. The computer uses mathematical algorithms to analyze the motion and predict the patient's intention. Then a signal goes to the motors in the hips and knees to move the trousers and the patient. Crutches are necessary, so the device is not feasible for people with paralysis of all four limbs. A rechargeable battery lasts for three hours, and the entire system weighs thirty-five pounds. Argo's website states that the device "uniquely matches the user's capabilities with the control mechanisms."[5]

Dr. Alberto Esquenazi, the chairman of the rehabilitation department at MossRehab in Philadelphia, was the principal investigator for a study of the efficacy and safety of the ReWalk. In theory, standing and walking would reduce the risk of complications associated with paralysis: pressure sores, osteoporosis, urinary infections, bowel dysfunction, pneumonia, and heart disease. Floyd filled out the application to try the electronic exoskeleton in March 2009.

Around Thanksgiving 2009, Floyd went through a screening process that included a bone-density test, X-rays, and an MRI scan.

"How was your bone density?" I asked him.

"They were surprised," Floyd said. "It looked almost normal."

"Thanks to all your sports," I said. "My most successful patients are active and athletic people."

Once Floyd passed the initial evaluation, he started a program of twenty-four sessions during eight weeks. He drove two hours each way from his home near Atlantic City to Philadelphia but considered this a minor inconvenience for the sake of using the ReWalk. His physician and therapists were impressed with his determination and progress. In fact, Floyd did so well with the device that ABC News featured him in a story for *Good Morning America* in early 2010.[6] (In December 2010, the ReWalk was featured on the television show *Glee*. In a miracle reminiscent of Tiny Tim's recovery in *A Christmas Carol*, a character on *Glee* named Artie used the ReWalk and crutches to rise up from his wheelchair and take a few steps, overcoming his paralysis. The electronic braces were a Christmas gift from an anonymous benefactor.[7])

When Floyd and I discussed his experience, I asked what he first thought about the ReWalk.

"Well, it's kind of strange looking," Floyd said. "I didn't really know what to expect. I'd never seen anyone else use it. These silver braces went up my legs to my hips, and the backpack was strapped on."

Therapists on both sides positioned Floyd as if he were about to stand up, and then he pushed a button to start the ReWalk.

"It started whirring and clanking," Floyd said. "Then the braces helped me stand. It took a few seconds to stabilize myself with the crutches."

As Floyd swayed to keep his balance, a therapist hit a button labeled WALK on the wrist control pad. As Floyd leaned forward, the electronic brace's sensors "realized" that he wanted to walk and sent a message to the computer, which activated the motors at his right lower leg.

"My right leg moved," Floyd said. "I couldn't believe it. Then the computer in the brace sent a message to my left leg. And my left leg moved."

With two strong physical therapists at his side, Floyd was walking—after three year of paralysis in his wheelchair, at the age of sixty-two.

"It was hard work!" Floyd said. "I was in good shape from all my athletics, but I needed a breather after a few steps. I sweated all the way, and sometimes the therapists had to keep me upright."

When he had walked up and down the hallway, a distance of seventy-five feet, Floyd returned to his wheelchair. A therapist pressed another button on the control pad, and the electronic brace let Floyd sit down.

"How did you feel after you walked?" I asked Floyd.

"I was so elated," Floyd said. "Standing up at six feet instead of being stuck at four feet in the wheelchair. I thought I'd never stand up again."

"That was quite a session," I said.

"Yeah, it sure was," Floyd said. "But it was tough. I wasn't used to doing that kind of work." His pulse and blood pressure rose, but not to dangerous levels, and his vital signs came back to normal soon after the electronic braces were taken off.

"You were a great athlete, so your body could handle that kind of work," I said.

"I was sweating a lot after the session, but it doesn't take a lot for me to sweat," Floyd said. "I couldn't wait to return the next day and do it all over again. I was so excited. I really can't explain it."

Completely paralyzed—with no hip, knee, or ankle function—Floyd needed both the structural stability *and* the motorized movement of the bionic braces. Linda Holmes had a similar problem after her stroke, but only with her knee joint (and she had paralysis only on one side). Imagine electronic legs on both sides, extending from the ankles to the hips, working in a coordinated manner. That gives an idea as to the challenges for engineers as well as paralyzed patients like Floyd (see figure 17.1).

After the national attention he received while walking with the exoskeleton, Floyd resumed his day-to-day existence with paralysis. Maryann had become a subspecialist nurse, learning more about spinal cord injury and its complications in the four years following Floyd's injury than in her entire career.

"It's been a lot harder on my wife than on me," Floyd said. "I think what she misses most is dancing."

"Where would you go for that?" I asked.

"There's a club called the Margate Log Cabin, about five blocks from our house. They've got a jukebox."

"What kind of music do Maryann and you like?

"My parents used to listen to country music," Floyd said. "I grew up on rock and roll, but we dance to singers like Marvin Gaye, Diana Ross, the Temptations."

"So what do you do for entertainment nowadays?"

"We don't get out much," Floyd said. "I get tired by nine and go to bed. We'll go out to the club for dinner once or twice a month. It overlooks the bay, nice view."

Whenever Floyd goes outside the renovated house, he worries about skin breakdown because he can't lie down as needed to reduce pressure areas. In spite of vigilant care, Floyd developed two pressure ulcers at his buttocks, and one area worsened after a coast-to-coast plane ride. He had a special cushion that evenly distributed his weight, but he was unable to use it on the airplane. Since January 2011, Floyd had been struggling with this pressure ulcer, which had forced him to limit his activities; he had already noticed a decline in his upper body strength. Floyd had developed some of the skin complications of paralysis even though he could use his upper body for pressure relief while seated; in contrast, Matt was tetraplegic and far more vulnerable.

Floyd stopped using the ReWalk after his role in the study ended. If he had been able to walk on a regular basis, he might not have developed skin problems, which can be never ending. He had plastic surgery for the pressure ulcer, but some stitches came apart. In the summer of 2011, he had to go back to the operating room. Floyd had to limit his mobility afterward, because even transferring from the bed to the wheelchair could cause shearing of the damaged tissues. He missed the fight against paralysis.

I wondered about resuming the conventional braces and asked him to compare them with the ReWalk.

"The regular braces were a lot harder to use, took a lot more energy," Floyd said. "I had to shift my hips to get just one step out of them. But with the ReWalk, every time I tried really hard, that messed up the movement."

"Yes, with the standard braces, you had to overwork your hips to move," I said. "That's very inefficient. The ReWalk imitates the normal pattern of walking. How long did it take you to figure out the difference?"

"About three weeks," Floyd said.

"The ReWalk senses what you want and helps you move. How did it sound?"

"Well, a lot of motors were running, whirring, and clicking. It's about half as loud as a lawn mower."

"Would I be able to hear it in an adjacent room?" I said.

"Maybe, maybe not. But the new one's much better; you can barely hear it in the same room. I saw it last fall at a conference in Las Vegas. And I met Argo's president, Amit Goffer."

"I hear he's working on a similar device, but for people with complete paralysis, like himself," I said. "What would you tell people with spinal cord injuries about the ReWalk?"

"Well, after a while, your body isn't that stiff. You can straighten up all the way. My wife thought there was better muscle tone in my legs. My general health was improving. I felt better as a whole."

"What about the other people in the study?"

"They felt about the same as I did. The first time you stand up you get this big old grin across your face. You don't have to say anything."

"It must be quite a feeling to go from a wheelchair to standing up and walking," I said. "The ReWalk is a remarkable invention."

"Yes, it is," Floyd said.

He stared at his legs, strapped to the titanium wheelchair, then at his muscular arms. A look came across his face as if he were imagining everywhere he would go with the electronic braces: the beach, a hiking trail, even the dance floor with Maryann.

18

In Flight

Matt, Early 2007

During a ceremony held at New England Sinai in February 2007, the hospital president, Tim Surgenor, and I praised Matt and presented a certificate: "Whereas you inspired the Cyberkinetics BrainGate Neural System Team through your courage and devotion during your participation in the clinical trial of the neural interface system, and whereas you have so capably and patiently represented the needs and hopes of those with paralysis to the world, Cyberkinetics NeuroTechnologies Inc. recognizes your efforts to improve the lives of others and extends the heartfelt gratitude and appreciation of the entire company."

After the event, I chatted with Matt and his parents.

"When I searched your name on the Internet, there were thousands of references," I said. "You brought a lot of attention to people with disabilities."

"If the BrainGate gave hope to people, it was all worth it," Matt said.

"I'm glad Cyberkinetics had this event," Ellen said as she gently rubbed her son's head.

"And the *Patriot Ledger* ran a good article," Pat said.

"Thanks for the comments about us, Doc," Matt said.

"You're welcome," I said. "They were sincere. The Nagles are good people."

"And thanks for the *Nature* magazine with your signatures," Pat said. "I tried to get one at Barnes and Noble, but they don't carry it."

"It's a major scientific journal, very specialized, very high level," I said.

"We got it from Harvard Square. Matt wanted to be on the cover of *Wired*," Pat said, referring to the March 2005 issue of the magazine in which Matt's story and the BrainGate were featured. "He was disappointed when he wasn't, but I settled him down."

"I was happy that *Nature* put me on the cover," Matt said.

"It's at least as good as getting on the cover of *Rolling Stone*," I said.

Later that year, John Donoghue was honored with the 2007 K. J. Zülch Prize for his "pioneering" research with "BrainGate, the mind-to-movement device that allows people with paralysis to control assistive devices using thoughts alone." Awarded by the Max Planck Society, this prize is "Germany's highest honor for basic neurological research."[1]

Matt's role in the BrainGate was over, and he moved on to his next quest: stem cells. Impatient with the pace of research in their own countries, patients traveled wherever they could get stem cell therapy. Matt heard about these people and dreamed of going abroad. One such person was Sonya Smith, an Australian who became paralyzed after a car accident. Two years later, she received eight weeks of embryonic stem cell therapy. Smith then claimed that she walked (with braces) and also regained "deep sensation" in her lower extremities and control of her bowel and bladder. She was among hundreds of patients who made a pilgrimage to India for embryonic stem cell treatment. Another patient had more sensation in his legs, less nerve pain, and better bladder function. He was pleased that his toes curled with stimulation of the sole, but I wondered if the stem cell doctor had informed him that this was only reflexive movement.[2]

In South Korea, scientists led by Seung Hwan Yoon injected bone-marrow stem cells harvested from patients into their own newly injured spinal cords. All had evaluations of movement, sensation, and nerve conduction before and after treatment. At four months, MRI scans showed healthy spinal cords and no bleeding, cysts, or infections. Muscle and sensory function improved in about 30 percent of the patients treated within eight weeks after the injury, but not after that period. The researchers planned to monitor all patients for long-term complications, including tumors and pain caused by nerve irritation.[3]

Matt's search for stem cells and a cure began to seem less urgent and less consuming, as if he was accepting his condition. Six years after Matt's injury and paralysis, Ellen began to notice changes in him. While they were sitting around, Matt often looked at his parents and said, "I love you, Mom. I love you, Dad." He called more frequently, especially during ballgames, to chat with his father about the merits of an athlete or a team.

"He's a different person," Ellen told me. At Shepherd, she had heard that it often took six or seven years to adjust to severe disabilities. That long? she had asked herself. Now she understood that the knife had transformed Matt into a completely different person, wandering in the strange country of tetraplegia and all its complications.

"Ellen's been fabulous through all of this," Pat said. "It takes a very special kind of person to deal with a child's catastrophic injury."

"You're a great team," I said.

"When I need a shoulder," Ellen said, "Pat's there. In his own quiet way, Mike's also been a great support. You've got to lean on each other. But it's hard."

"I try to get up each day with a smile. I try to enjoy life," Pat said. "I'd drive myself crazy if I was having coffee and thought that Matt couldn't unless someone held it for him. I just can't beat myself up. If I can make Ellen and myself happy in some small way, then we're staying healthy to help Matt."

"Yes," I said. "Take care of yourselves so that you can take care of Matt."

Matt's fits of anger and bitterness about his condition had subsided, and he learned to extinguish smoldering conflicts with people.

"He's more giving," Ellen said. "He gets joy from the lives of others, as though he lives through them. He wants the details, like where we had dinner and what we ate."

"Matt seems more empathetic," I said.

"Yes. We saw a news story about a baby who died in a car accident, and Matt said, 'Mom, that breaks my heart.' I didn't say that he broke *my* heart because he felt *so* much for others even though he's paralyzed in a wheelchair."

"He still has you and Pat and Mike," I said.

"We had the best talk last Sunday." Ellen smiled. "A wonderful visit. He was so grateful for all we'd done. Matt said that nothing could express how much he loved his family."

"A couple of Sundays ago, Mike and he had dinner," Pat said. "They rarely get that one-on-one time. They've had some trouble."

"Trouble?" I asked.

"They're so different," Ellen said. "Mike was angry about the way Matt acted. Mike's a quiet person, and he didn't like Matt's crowd."

"Mike doesn't know how to handle Matt and vice versa," Pat said. "Matt could do the worst things in the world . . . but he made you laugh, and you'd forgive him."

"They've been connecting more and more," Ellen said. "Matt always says, 'Tell Mike I love him.' He told Mike to start a construction business because he was so talented. If he could build this place for Matt . . ." Ellen gestured with both hands at the dining room with French doors leading out to the wooden deck.

Matt and I talked on occasion about stem cells. We were both excited to hear of a breakthrough in 2007: Rudolf Jaenisch of MIT, Shinya Yamanaka of Kyoto University, and Konrad Hochedlinger at Harvard University transformed *adult* mouse skin cells into cells that acted like *embryonic stem cells*.[4] Researchers isolated four genes—specific sequences of DNA—that make cells behave like stem cells, turning back time in the genetic clock of cells. These genes were spliced into retroviruses, messengers that inserted the genes into the DNA of the skin cells. Now the genetically engineered cells could form various tissues—just like stem cells. But this was a cautionary miracle. Even though the viruses were modified to prevent cancers, mice that were cloned from the engineered skin cells developed tumors.

If human adult cells were reprogrammed into embryonic stem cells, then embryos or eggs would not be required to create therapeutic stem cells. In fact, later that year, Yamanaka achieved the same feat with *human adult* skin cells.[5] Richard Doerflinger, a spokesman for the US Conference of Catholic Bishops, said that this research "does not raise the moral problem of creating or destroying embryos; it may offer a way for people of all faiths and all ethical backgrounds to study, use, subsidize, and enjoy any therapeutic benefits of . . . stem cell research."[6]

Now patients could potentially grow their own organs without risking immune rejection, because the cells had their own genetic structure.

"These are enormously important papers," George Q. Daley, president of the International Society for Stem Cell Research, said.[7] This research avoided ethical issues and was eligible for federal funding. "It's a gift of God," a priest told me.

Although Matt and I were optimistic about these programmed cells, they had not been used to treat spinal cord injuries. Then Freda Miller and other researchers at Toronto's Hospital for Sick Children injected stem cells derived from skin cells to treat spinal injuries in rats.[8] Acting like embryonic cells, they covered nerves with myelin and sped up electrical transmission. The implanted cells created a bridge across the injury, helped nerves to regrow, and improved coordination as well as the ability to walk. This technique had both ethical and clinical advantages. Skin cells were genetically reprogrammed to generate stem cells—without using embryos. Also, if the stem cells were derived from patients with spinal cord injuries, there would be almost no risk of an immune rejection. But it remains unclear whether genetic engineering would make these cells cancerous.

<hr />

Ellen Nagle paged me on Thursday, July 19, 2007. From her tearful voice, I knew something was terribly wrong. Matt was in Good Samaritan Medical Center in Brockton with a very bad infection that had sunk his blood pressure and made him unresponsive. Now he was comatose and only blinked on occasion.

"Matt was over on Sunday. We smelled a bowel movement, but he didn't know about it. Matt was coming down with a bug. Pat wanted to get him back to Sinai, but I had made his favorite shish kebabs. Matt had brought a stack of DVDs—*Spider-Man*, other action movies." Ellen's words stopped rushing out, and she blew her nose. "We never saw them."

Ellen knew that people with spinal cord injuries often have infections, so she had learned to take them in stride.

"We took him back. Danny went to see him the next day, and he still wasn't feeling well. The nurse started IV fluids. Matt seemed OK, but he was confused about the IV in his hand. Danny told him he needed the fluids for his blood pressure."

That was a bad sign, but I said nothing to Ellen. Around midnight, even though the IV machine pumped fluid into Matt's veins, his blood pressure dropped, and the nurse couldn't even get a reading. Severely hypotensive, Matt was given intravenous dopamine and rushed to the ER, where he was diagnosed with septic shock and wheeled into the ICU. Possible sources of the infection were a bedsore on his back, his lungs (his sputum looked infectious), and a urinary infection. Matt also had some infectious drainage at his toes due to an injury, a consequence of not having any sensation below the shoulders.

"I'm so glad Danny was there," Ellen said. "I hope Matt just nodded off."

"That's probably what happened," I said. "As Matt became feverish and his blood pressure went down, he would've felt tired and lethargic. It's not painful."

"He wouldn't have been aware of anything, though?" Ellen said.

"No. No, he wouldn't."

"He wouldn't feel himself suffocating, would he?" Ellen said.

"No," I said. "It's like going off into a deep sleep."

"That's a comfort."

His CT scans from Tuesday and Wednesday appeared normal, so I said, "That's very good news. By now we'd see brain damage if there was any."

I asked Ellen about Matt's kidney function and was relieved to hear of his good urine output and normal labs.

"That's also a good sign," I said. "Ellen, don't give up hope."

I tried to be optimistic. Matt was young, and I had treated older patients with hypotension due to sepsis who had done reasonably well.

That evening, I called the ICU. Matt's nurse said that his labs and blood pressure were fine and that he was generally improving, except for his brain function—he was still comatose. Matt was on tube feedings, with an insulin drip because of his diabetes. His chest X-ray showed either pneumonia or partial collapse of his lung—or both.

"His brain scans were fine," the nurse said, "but he's not waking up."

There was no kidney failure, so I held out some hope that his brain had also been spared. And he had received steroids to reduce brain swelling. Perhaps a drug for septic shock—Xigris—would work its magic.

On Friday, the Nagles told me Matt had declined. He didn't blink in response to conversation or to signify "yes" with his eyes. But his blood pressure was normal, his ventilation was fine, and his labs had also improved. In response to the infection raging through his body, Matt's white cells had skyrocketed to fifty-one thousand, but now they were just above ten thousand, which was almost normal.

I tried to be as encouraging as possible and kept my pessimism to myself. Matt's brain had clearly suffered some damage. I wondered if subtle cerebral swelling had been present on the first CT scan but was now worsening.

On Saturday morning, I drove north on Interstate 95 toward Boston. It was sunny when I set out, the clear blue sky scarred by a few cirrus clouds, the highway lined with pines, oaks, and maples, a perfect summer day for a weekend drive—except for my purpose.

Good Samaritan Medical Center, a tall rectangular building topped by a cross, is ensconced in a wooded area, its entrance flanked by two clusters of juniper bushes. In the ICU waiting room, I gave Ellen a hug and shook hands with the Nagle men. Pat was proudly wearing the Cyberkinetics T-shirt from Matt's event. We sat down, and they gave me an update. Matt had just been sent for an EEG and another CT scan.

"So, what do you think, Dr. Mukand?" Ellen's hands were clasped as if in prayer.

I had to give them an infusion of hope.

"Matt's kidneys are doing well, in spite of the drop in blood pressure. Kidney and brain damage go together, so I'm hopeful that his brain is better than it seems."

"But why isn't he talking?" Ellen said. "His eyes just roll up. It's so sad to see him like that. Someone said he might have had a stroke?"

"What exactly does that mean, Doc?" Pat said. "I've seen people with strokes and they're usually weak on one side. . . ."

"Well," I said, "if Matt's blood pressure dropped, that could damage the brain. Areas called the watershed zones are most vulnerable. After brain cells die, there's some inflammation, and . . ." I looked at their sorrowful faces and hastily added, "But it's too early to say if that's what happened with Matt."

Ellen sighed, and there was a long silence. Her hands were still clasped together as she leaned forward, almost in a fetal position. At the other end of the floral couch, Pat slumped down, his arm hanging limply over the armrest. I withheld the sad details of encephalopathy, which could have affected the blood supply to Matt's brain.

Ever since 2001, July had been the worst month for the Nagles. America's independence celebration was always linked to Matt's paralysis. They had planned to see the Pawtucket Red Sox that weekend. Instead, they were huddled in a waiting room.

Ellen asked me to meet Matt's doctor, perhaps hoping that a physician would receive some clarity. Through a phone intercom, the Nagles and I announced ourselves, cleaned our hands with waterless antiseptic, and entered the ICU through heavy swinging doors. Ellen introduced me as a family friend and Matt's BrainGate doctor. She had brought along a copy of the *Brown Alumni Magazine*. His nurse, Marsha, took one look at Matt's photo and said, "Wow."

"Wasn't he handsome?" Ellen said.

"He still is," the nurse in green scrubs said.

In a small conference room, the Nagles, Dr. Pahuja, and I reviewed each sad page of Matt's record. The careful attention to Matt's many medical problems was impressive.

"His labs and vitals normalized quickly," I told Ellen and Pat, wanting them to know that Matt had received excellent treatment, to be relieved of any painful doubts about his care. "The ICU did a great job with Matt."

"Thank you," Dr. Pahuja said quietly after the Nagles had left. We both knew that the best treatment in the world might not be enough. "But I wish I knew what was happening with his brain. Maybe he had a brain stem infarct."

"Yes, it could be the reticular activating system," I said, referring to the nerve pathways responsible for alertness and the sleep-wake cycle.

Dr. Pahuja sighed, his eyes intense behind wire-rimmed glasses.

"Spinal cord injuries lead to all kinds of complications," I said. "Even small pressure ulcers can get infected and poison the whole body."

"Sepsis is hard to fight," he replied.

"When do we get the EEG results?" I said. "It was done a couple of hours ago."

"I'll find out. And the CT should be done soon. Are you here for a while?"

"Until we get some answers about his brain."

I wondered just how long and how severely Matt's brain had been deprived of blood and oxygen.

In the waiting room, the Nagles and I were barely seated before the questions began.

"What do you think?" Ellen's blue eyes brimmed with hope.

"They've done all the right things," I said, nodding for emphasis. "But I don't know how long Matt was hypotensive. . . ."

"I saw him yawning a while ago. That's a good sign, isn't it?" Ellen asked.

"It's a reflex," I said softly. How could I tell them he was at a very low neurologic level?

"Did he say anything about a brain stem stroke?" Pat asked, gripping the armrest.

"What's the brain stem?" Mike asked, desperate to understand what was happening with his younger brother.

"The brain stem controls the lungs, the heart, and the level of consciousness." I knew that damage in this area meant a harsh prognosis, but I held my tongue. Who was I to take away their hope?

On Matt's return from the CT scan, we again washed our hands with waterless antiseptic and went to his room. Ellen and Pat wanted me to examine Matt.

Comatose, with his eyes deviated toward the right and upward, Matt could not respond to any of us. He was pale from the lack of sunlight and his anemia; he rested on a bed with an air mattress to prevent pressure ulcers. A computer screen for his Puritan ventilator showed the inspiratory and expiratory waves of his breathing through blue plastic hoses. The machine forced oxygen into Matt's lungs, hyperventilating him twenty-four times per minute to reduce his brain swelling. The Space Lab monitor showed a heart rate of fifty beats per minute, a blood pressure of 128 over 81, and a central venous pressure of twelve millimeters. Matt still had the heart and lungs of an athlete.

Positioned on two pillows, Matt's arms were swollen with fluid from lack of use and range of motion. White plastic sleeves connected to a VenaFlow system compressed his legs to prevent blood clots. Three pumps stood at his bedside—one for tube feedings, one for the Xigris drip, and the third for insulin and the antibiotic ceftazidime.

"Matt, how're you doing?" I said loudly, hoping he could hear me. "I'm very sorry you're in here, but we'll get you out of this place. Hey, the Red Sox won yesterday. And what's more, the Yankees lost. Wouldn't it be incredible if we won two World Series in our lifetime?"

The Nagles smiled. Hearing about his beloved Red Sox would certainly get a response from Matt. But he kept staring up at the ceiling.

To assess his brain function, I did some simple tests. Both pupils responded to light by constricting. Then I tried the "doll's eye" test, to check a pathway in his brain stem. When the head is rotated to one side, the eyes tend to look in the opposite direction to maintain the gaze. I didn't think that Matt's response was entirely normal, but I didn't say anything. Matt didn't protectively blink when I quickly moved my hand toward his eyes. He had no sensation below his shoulders, so I pinched his neck to give a painful stimulus. No response.

Matt was back in the ICU, where his journey had started six years ago. His ragged and subdued brain waves on the EEG did not make me feel optimistic. Dr. Pahuja and I went to review the CT scan. A kind radiologist heard of my roles as friend and researcher and was most helpful. On the computer screen, he scrolled through the CT scan slices of Matt's brain and pulled up the previous study for comparison. His eyes narrowed as he looked at us for more clinical information.

We explained the severe hypotension and the lack of brain circulation that occurred for who knew how long. Then the radiologist pointed out abnormalities in the basal ganglia, the watershed areas that I had mentioned to the Nagles. On both sides of his brain, Matt had lost the normal differentiation between the gray and white matter in the temporal and parietal regions; the crevices between the contours of brain tissue were also gone. This abnormal brain landscape meant that he had severe brain swelling.

"He's headed toward brain death." The radiologist stared at us for emphasis. "Did he have an EEG?"

"That didn't look good either," Dr. Pahuja said. "Not much activity."

"Well, I guess that's it," the radiologist said.

On the other CT scans, Matt's lungs were inflamed on both sides due to pneumonia. And his abdominal scan was abnormal—his liver and spleen were enlarged, and his right kidney had stones. But the left kidney had been spared. A Foley catheter for his bladder and a rectal tube were also visible.

On the way to the ICU, Dr. Pahuja and I discussed how to break the news.

"Please take the lead." I didn't want to interfere with the ICU team. "I'll back you up."

"I'll be honest, but I won't take away all their hope," Dr. Pahuja said.

In the family waiting room, we learned that Pat had taken Mike back home. Unable to tolerate what was happening to his brother, he was at the point of breaking down.

Dr. Pahuja delivered the CT scan report as kindly as any physician could. Then I sat down with Ellen and her questions:

"What did he mean by brain damage?"

"Will Matt get better?"

"What about the EEG?"

"What exactly did the CT scan show?"

Cell death had occurred in multiple regions of the brain. I had already told the Nagles how hypotension causes inflammation, swelling, and cell death, especially in the watershed areas. Now I had to be more clear.

"This isn't Matt." Ellen started crying, and I placed my arm around her as we sat on the couch. "When he first got injured, I could handle it. At least he was still Matt. He still had his sense of humor; he still got angry. But this isn't Matt." She wiped away her tears. "Is there any hope?"

I described some of my other patients with hypotension. One had an excellent recovery. Another was confused and weak. A third never made it out of the ICU.

"It's still early. And he's young, and a fighter. Look at what he's been through already."

Ellen nodded, smiled weakly, and collapsed back into the couch. I offered to bring her dinner because it was getting late, but she wasn't hungry. She thanked me for coming, and I told her to call me at any time.

In the lobby, I walked through the unremitting sunlight, and the glass doors slid open. On the front wall was a cross. Ellen had said, "I'm glad he's in a Catholic hospital so the priest can come every day."

On Sunday, July 22, while picking blueberries with Jacob in the afternoon, I called the ICU. Matt was still comatose. His EEG waves were slow and silent, almost flat. A blood-flow study, in which radioactive material was injected inside Matt's brain, showed only "minimal activity" in one area and "no activity in the rest of the brain."

I had planned to let Ellen and Pat keep a few milligrams of optimism until Matt's neurologist discussed the CT scan, the EEG, and the blood-flow results—and gave a poor prognosis. But the Nagles already knew they had lost their son. Their parish priest, Father Sean, had given Matt the Anointing of the Sick.

Another blood-flow test on Monday was "consistent with brain death." When I called early that evening, a nurse told me in a solemn voice that I should speak to Dr. John Keefe. He answered my questions clearly and kindly, and I knew we had lost Matt. His brain was severely damaged.

"We can arrange an OR spot for his organs," Dr. Keefe said in a hushed voice. The Nagles wanted to honor Matt's wish to donate his organs. "I'm surprised he still has pupillary responses."

I was silent for a moment. Technically, the pupils should not constrict with light after brain death. But he was almost there.

"That's Matt, fighting till the end."

"His family wants only comfort measures. We'll take him off the vent in the morning. Don't worry."

"Thank you." I paused and tried to control my voice as I dabbed at my tears. "The ICU staff was great," I said. "I reviewed the record with Dr. Pahuja. You guys did everything, everything. . . ."

Dr. Keefe thanked me and handed the phone to Ellen, who said, "We have to let him go. Matt wanted his organs to help someone who's ill."

"He was a generous person," I said. "It's his last gift."

Through her sobs, Ellen said, "I can see him on the five-yard line, offering his heart." Then she said Matt had thought highly of me. "He was tough at times, but I think you understood him."

"I tried, I tried," I said. Then I could say no more.

"Matt wanted to be cremated," Ellen said, "because he'd get claustrophobic in a coffin!"

"That's Matt," I said. "Joking in the face of death."

On my way home from my rehabilitation center, I drove through a gray mist that shrouded the car. Over dinner, I said bitterly to my wife that the stabber may as well have killed Matt with a clean blow instead of making him suffer for the past six years.

"But then his parents wouldn't have had him all those years," Giselle said. "And he and you wouldn't have accomplished what you did."

While I drove from Providence to the ICU in Massachusetts, the mist obscured my vision, but even in the hazy light, the tall pines were lush and strong. At the roadside, a doe materialized, waiting for my car to hurtle past. During the ride, I wondered what to say to the Nagles. How could I comfort them after all they had endured in the six years since Matt was stabbed? I remembered the times when Matt and I had worked together. We had argued sometimes, but mostly we had achieved our goals for the BrainGate and for people with disabilities. I remembered when we had worried about the Red Sox, joked about his meeting Britney Spears, discussed the next phase of the BrainGate, and debated about stem cells in Korea.

In the ICU waiting room, Ellen, her sister Colleen, and a cousin sat near a vending machine, a half-empty carafe of coffee on the table. With a resigned flicker of a smile, Ellen greeted me. She also introduced me to Dan Miller—from the New England Organ Bank. Soon everyone's vigil would be over.

"The last six years have been incredibly intense," I said.

"Yes, that's right," Ellen said.

"You were dealing with the worst emotions—grief, fear, bitterness, anger."

"Yes. And the loss of what could have been," Ellen said. "The hardest was when Matt got hurt, but I've processed that. I'm devastated that I'll never see him again here. But I know that he's at peace."

"You had that wonderful weekend recently, sitting under the oak tree."

"There was a gentle wind, and it was sunny," Ellen said. "We talked about a movie that Matt loved, *Forrest Gump*. It started with the paper bag and the feather, and the girl who just wanted to fly free. Matt said that he wanted to fly free, just like the feather and the bird."

"He was weighted down in a terrible way," I said. "Matt told me that one of his favorite songs was by the Steve Miller Band. The singer wants to fly like an eagle to the sea."

"When they told us about Matt's brain damage, I said that he was in flight," Ellen said. "Father Sean prayed with us. We had to tell Matt it was OK to go, because he was hanging on for us. So we told Matt he'd be with NayNay and Papa. I swear to God, his eyes moved, and there was a little tear. If a priest weren't a witness, you wouldn't believe me. Then Father Sean sat with us and told Irish stories, funny ones that made us laugh. It was like an Irish wake."

According to Irish custom, a window was opened for Matt's spirit to fly away.

"It's fitting that Father Sean came," Ellen said. "He'll do Matt's service. It's like the circle is complete, because he had talked to Matt and Mike after the injury."

We chatted about Matt's life, everyone he had touched, and a few choice escapades. Then I went to see Matt for the last time. I did not open his eyelids to see if he still stared up at the ceiling. Nor did I test his pupils or corneas or eye movements. I didn't check if his arms were abnormally positioned due to having lost almost all brain function. Matt's neurological examination revealed one diagnosis: peace.

For the last time, I touched his shoulder and said, "Good-bye, Matt. Take care." Then I cleaned my hands with the waterless antiseptic, slowly walked out to the waiting room, and gave Ellen a hug to help her through this last phase as Matt's mother.

It was eleven at night when I left Good Samaritan Medical Center. While I drove back to Providence, a fine mist still hovered above the windshield, but visibility had improved. On the car stereo, I listened to a CD of the veena, a stringed Indian instrument with a sinuous and sonorous sound. Its dark notes were a fitting dirge for a young man. I drove on, my eyes fixed on the distant horizon.

On July 24, 2007, Matt was wheeled down to the operating room for the last time in his life. Then he was disconnected from the ventilator. His athlete's heart, strengthened by years of skating on the ice and running the fields and exploding on the defensive line—the heart that had faithfully pumped for six years in his electric chair—trembled as the cardiac muscles lost their oxygen. On the monitor, the regular, spiked green waves became chaotic, then ebbed away into a quivering line. His heart stopped. Soon, the surgeon pierced Matt's body with a flexible tube that had a blade tip. He snaked it into the aorta, the body's largest artery. Matt's crimson blood was drained and replaced with a clear, ice-cold preservative. Through strategic incisions, the surgeon poured a cold salt solution over Matt's organs to slow down their metabolism; then he preserved them on ice. Now Matt was a donor, a gift to anyone who could benefit from his organs and tissues.

Aflame in Red Chrysanthemums

Matt, July 2007

O n a verdant summer day, I drove through Weymouth to McDonald
Funeral Home, a large white colonial building with pillars holding
up a spacious porch. Police cars and motorcycles from all over Massachusetts were lined up in Matt's honor. Uniformed officers stood at the
entrance as if to protect Matt.

Inside the small lobby of the brick funeral home stood a table with a
flower arrangement of tiger lilies, yellow roses, and sunflowers sent by
the MIT police. After signing my name in the guest book, I stepped into
the room where Matt lay surrounded by flowers. His face was dispas-
sionate and pale. In the past, I would receive a wink if I said something
even remotely funny or if we came to an understanding about his medical condition. Now he would not ask me to dial his phone or give him
one last drink with a straw before I left his room. Near his head, a bou-
quet was aflame with red chrysanthemums and orange birds of paradise.
Matt's hands looked more pale and puffy than usual, and he appeared to
grasp a rosary woven through his fingers.

The Nagles greeted me, and Ellen gestured at the adjoining room,
where a gallery of photographs included an enlargement of Matt on the

cover of *Nature*. In one photograph, Matt wore his Boston Elites hockey shirt while standing near a large cake in celebration of a victory. A photo taken after his injury showed him grinning away, connected to the ventilator via plastic tubes, a cousin with his baby son at his side (see figure 19.1). Matt loved children, and they responded to his mischievous smile and unpredictable sense of humor.

While we mourners sat down for the service, Eva Cassidy, the talented singer who had died of cancer at thirty-three, sang over the intercom. She reassured us that somewhere over the rainbow, with the clouds far below, one could fly through the clear skies with bluebirds, and dreams would come true.

During the short service, a boy in a motorized wheelchair started having leg spasms. I approached him and whispered an offer to stop them. As I knelt down and repositioned his legs to stop the spasticity, he nodded and smiled his thanks. I patted his shoulder.

Each of us took a flower and followed a motorcade of police cars and motorcycles that escorted Matt in a hearse to Saint Francis Xavier's Church. Our procession wended through the town of Weymouth, past neatly manicured houses with stone walls, gas stations, stores, the parish center of Immaculate Conception, the Weymouth Police Department, and the Massachusetts Electric building while police officers on motorcycles with flashing lights guarded our route. The cross traffic waited quietly, except for a truck that almost barged into our cavalcade but stopped short to let us proceed without interruption. As we drove past Weymouth High School, I saw the football field and imagined Matt galloping across the ten-yard line in pursuit of a running back.

Saint Francis Xavier's Church was a large but modest brick building adorned with stained-glass windows, a relief statue of Jesus, and a large pillar capped by a cross.

Lord, have mercy.[1]

At the entrance, priests in white robes welcomed us into a lobby that held two marble receptacles for purified water.

Lord, we have sinned against you.

Tall, colored windows softened the summer sun and helped keep the church cool. Instead of extravagant chandeliers, the ceiling had light fixtures with a cross at each corner.

The Lord be with you. And also with you.

Small electric fans whirred quietly overhead, losing their battle against the heat as they tried to cool down the seven hundred people in the church.

The eagle shall be your chariot to the Lord, and you shall rise up through the clouds into his eternal shelter, where the sunlight shall bathe you in his warmth.

A congregant marched in with a cross to start the service. Matt's coffin was wheeled in, covered with purple velvet. Three tall candelabra flanked the tabernacle, its white marble triangular base truncated by a black marble slab and inlaid with golden fish.

By suffering on the cross He freed us from unending death, and by rising from the dead He gave us eternal life.

In *Celebrating the Eucharist*, the gospel reading for July 28, 2007, was Matthew 13: 24–30: A man sowed healthy wheat seed, but an enemy added a poisonous rye grass, which became apparent when the plants sprang up. Realizing that this was an enemy's work, the man resisted the urge to weed out the poison grass and, in the process, uproot the wheat. When the wheat and the grass matured, he asked his reapers to pull out and burn the grass, then harvest the wheat for storage in his heavenly barn. It was a good crop, one that nourished many people for a long time.

We offer this holy and perfect sacrifice: the bread of life and the cup of eternal salvation.

During the service, I envisioned Matt's organs at the New England Organ Bank, stored under antiseptic conditions in special electrolyte solutions, waiting for a proper match of his blood and tissues to patients who could receive his gifts. His left kidney was already on its way to a nurse in New England who had kidney failure due to lupus.

I am the Lord of the earth and the wind and the fire and the water, and I will heal the sick and the lame, for I have heard them suffer in their darkest nights, and I shall bless them when they cry out my name.

High on the white marble front wall were relief statues of Jesus and Mary, made of red stained glass. At one side a man supplicated to Jesus, and on the other a woman climbed a stairway.

I shall scatter your darkness and your fear with a sweep of my hand. I shall give sound to the deaf and light to the blind and movement to the paralyzed.

A priest walked around Matt's coffin with a silver incense holder suspended by silver chains. From the censer, a smoky aroma rose up to

God, and the congregants chanted, "Let my prayer arise in Thy sight as incense."

All who embrace the Lord shall forever find peace and life.

At the pulpit, Mike Nagle read from the fourth chapter in the Book of Wisdom: "The just man, though he die early, shall be at rest. For the age that is honorable comes not with the passing of time, nor can it be measured in terms of years. Having become perfect in a short while, he reached the fullness of a long career; for his soul was pleasing to the Lord, therefore He sped him out of the midst of wickedness."

Let us proclaim the mystery of faith: When we eat this bread and drink this cup, we proclaim your death, Lord Jesus, until you come in glory.

With a final blessing and a prayer, the priest waved the silver incense burner around Matt's coffin at the end of the service.

In the Lord there is the peace of dawn.

In the Lord there is the calm of dusk.

I imagined eight-year-old Matt as an altar boy in a black cassock, covered by a starched, white surplice as he solemnly extinguished the votive candles.

In the Lord there is the harmony of the trees in the wind.

Priests stood at the church doorway, shaking hands with the hundreds of churchgoers as they left and giving each a white balloon.

Through Him, with Him, in Him, in the unity of the Holy Spirit, all glory and honor is yours, almighty Father, forever and ever.

When Matt first got injured, he had said to Ellen, "Do you think I'll have a big funeral?" She told him not to talk like that.

The peace of the Lord be with you always. And also with you.

Father Sean said that the balloons symbolized the ascent of Matt's soul to heaven. Each of us released a white balloon that the summer wind swept up higher and higher toward the blue heaven until the white specks vanished.

May almighty God bless you, the Father, and the Son, and the Holy Spirit. Amen.

We slowly brought our eyes back to earth and walked in silence to our cars.

Go in peace to love and serve the Lord.

Following the service, we drove to the Knights of Columbus hall, a large building where the lunch reception would be held. Its front lawn was graced by two clusters of yellow lilies. During the event, Ellen and Pat introduced me to Matt's extended family and friends. Father Sean thanked me for taking good care of Matt and said, "God bless you."

Family members, friends, and I had advised the Nagles to get away after the funeral. Recently back from a trip to Aruba, Ellen and Pat invited me over for lunch in the fall of 2007. Ellen prepared her specialty, which was also Matt's favorite meal: shish kebabs. We ate in the beautiful room that Mike had built, its cherry floors gleaming in the sunlight that streamed through the windows with red and white floral curtains. As we chatted, the questions became more serious.

The Nagles asked about the cause of Matt's death. I agreed with Dr. Bloom that Matt had developed resistant bacteria after multiple urinary infections and repeated antibiotics. Even though Matt was given antibiotics, they may not have been strong enough to quickly kill a potent new strain.

We talked about their trip for a while; then Ellen sprang another difficult question.

"As a rehab doctor," Ellen said, "what was your impression when you saw Matt—when he wasn't awake but his eyes were open? Did you believe there was a level of consciousness? Honestly, I'd really like to know what you think."

"Based on the EEG report, I didn't think there was much consciousness." I didn't think—and didn't want the Nagles to think—that Matt had experienced any pain.

"Even in the beginning?" Ellen frowned.

"Well, from the time that the EEG was done." She had pinned me down, and I had to be more careful.

"What did you think of his prognosis at first?"

I looked over at Pat, who nodded his head slightly.

"When I first saw Matt, I wasn't very optimistic. But he was young, so that gave me a little hope. The CT scan made me pessimistic."

"Matt bounced back so many times," Pat said. "I thought he might do it again. Except in the last two days."

"Why were his eyes open sometimes and closed at others?" Ellen asked.

"I think there were times he seemed to know what was happening," Pat said.

"He may have been going in and out of consciousness," I said. "And then the brain damage got worse, and he became comatose."

"He was in bad shape." Ellen's hand shook slightly as she handed me the grilled vegetables. "When I read between the lines, I realize it was worse than we thought."

"Matt would've woken up with Ellen's poking and prodding, fixing his hair and grooming him." Pat smiled and said, "He would've told her to leave him alone."

"You wonder if they had found him sooner . . . ," Ellen said.

I was afraid of exactly this line of thinking—an obsession over possibilities.

"You can go crazy with what-ifs," Pat said.

"You can kill yourself with what-ifs," Ellen agreed.

"Yes," I said. "You could go back six years and deal with all *those* what-ifs."

"I took comfort in knowing that Matt would be at peace, no longer suffering," Ellen said. "So we told Matt it was OK for him to go. But you know, selfishly, I wanted him. I never wanted him to go."

"You let Matt go because he was ready to go," I said softly. "Father Sean gave the blessings. His soul was gliding away."

All three of us were silent for a while. I thought about how, when Matt was cremated, the metal in his neck melted away. He was no longer at knifepoint.

"Someone told us she had cared for her quadriplegic child for twenty years," Ellen said. "I really believed Matt would outlive us."

"And Mike had to be ready; he knew that," Pat said. "When we made our wills, the lawyer said we should set up a trust. We couldn't put anything in Matt's name."

"At Shepherd, they said he'd probably have a long life," Ellen said.

"It really varies," I said. "Look at Christopher Reeve."

"That's what really scared me," Pat said. "Look at all the money and the people he had, and he lived only nine years."

"I think the diabetes was really a bad thing," Ellen said. "He had so many problems; nothing was ever easy. We were lucky to have him for six years."

"I would've lost Ellen and Mike if he hadn't stayed alive that night," Pat said. "I definitely would've lost her. And Mike was so wild I don't know what he would've done. When it happened, I went out with my gun, and my guys had to stop me. Who knows what I would've done if Matt had died."

"Well, it's nice you're here because it's our first Sunday home after our trip," Ellen said. "And we would've been alone."

"Yes, I was afraid I wouldn't get nice meals on Sundays anymore," Pat said.

"I'm glad to be your excuse for a good meal," I said to Pat. I finished a bite of my shish kebab. "Now I know why this was Matt's favorite meal."

"Matt loved good food. And he also liked a good laugh."

Pat described one of Matt's favorite TV characters, Fire Marshal Bill, a caricature played by Jim Carrey. One summer day, while the Nagles sat in their cabin at Lake Winnipesaukee, watching the rain, Matt did his impression of the fire-safety officer, making up his own skit. The lanky teenager in a New England Patriots shirt projected his front teeth over his lower lip and strutted around the room, his hair disheveled like that of the comic fire marshal, a crazed fire in his eyes.

"Now," he said, "here's a common danger at camp: the tape player. Sure, it makes music, lots of fun. Let's say you're sitting around, listening to the music, but the battery dies. So you plug it in. But you're standing in a bucket of fish bait."

The fire marshal plugged the tape player into an imaginary electrical outlet, started shaking as if electrocuted, and fell to the ground, quivering in shock.

"Are you OK?" his family asked in unison, having seen the TV version of this skit.

Fire Marshal Matt stood up, brushed his hair out of his eyes, smiled his toothy smile, and said, "Never better!"

The three of us chatted about Matt's sense of humor until Ellen suddenly went off to the kitchen. Then Pat and I quietly stared out the French doors at the sunlight on the pine deck.

"Yesterday, we took the Corvette down to Plymouth for an ice cream," Pat said. "I told Matt if he ever got stem cells and could move again, that car would be his. Now I think about him in a little two-seater with no protection. Am I insane?"

He laughed at himself.

"I always believed that one day Matt would walk, no matter what anyone said. It wasn't just a fantasy. If anyone could make it happen, Matt could."

20

Passing the Baton in the
Race with Paralysis

At Mount Hope Cemetery, Ellen sat on the black granite bench engraved in silver with Matthew Nagle's name. Nearby were the American flag and a Red Sox World Series pennant. She remembered cheering Matt during the Thanksgiving football game that had made him famous and drawn letters from colleges that might have led to a different life. She opened the sports pages of the *Boston Herald* and began reading the news to Matt; she imagined him grinning at her, his blue eyes bright with a mischievous comment that would soon bubble out of him.

‖‖‖‖‖‖‖‖‖‖‖‖‖‖‖‖‖‖‖‖‖‖‖‖‖‖‖‖

Pat walked along Rustic Drive on his way to the cemetery, holding a brush and a football. He cleaned off the granite slab where his son's cremated remains, free of the knifepoint, were buried. As for the football, he wasn't sure if he would lay it down in the flowered end zone or sit with it on the bench or toss it at a pine tree, far away from the graves and monuments, and wait for a skinny fifty-pound boy with weights in his pockets to chase after the ball and bring it back to his father, daring Pat to tackle him and laughing with triumph. When Pat dreamed about Matt, the same image burst into life: a five-year-old boy with front teeth missing, laughing at his father, his family, his friends, and the world.

Danny Taylor and I strolled along Wessagusset Beach.

"He was still my buddy, no matter what. He was still my buddy."

He paused to pick up a shell, blew off the sand to reveal the iridescent interior, and watched the particles fly off in the wind.

"He had so many problems after being stabbed. It was my place to help him out."

I placed my hand on his back and said, "Matt thought the world of you."

Danny stared past the beachgoers, past the gently rippling tides of the inlet, past the sailboats that kept their sails unfurled, into the infinite horizon.

"A lot of free time now that he's gone. Sundays aren't the same."

Katie Perette flipped through the collection of news clippings and posters that she had sent to the New England Patriots and the band Aerosmith for Matt's fund-raiser. On the cover was a photograph of a handsome blue-eyed young man with a plastic tube in his throat and a smile that refused to be suppressed. Next to him was a young, attractive blond woman: Katie. What if the fireworks hadn't lit up the sky and the music hadn't quickened the heartbeats of the beachgoers into a fight? What if the stabber hadn't been intoxicated with violence and the sunset hadn't bled through the clouds?

Matt and I had planned a trip to his favorite restaurant, Monponsett Inn, on the edge of a lake in Plymouth County. Sitting on the pine deck, we would have looked out across the lawn, past the white petunias and orange snapdragons, toward the water glistening in the sunlight. We would have chatted about the Red Sox and the highly improbable odds of their winning another World Series. He would have felt the benediction of the wind and the warm hand of the sun upon his face. I would have tilted a bottle of Sam Adams Summer Ale for Matt, a man young enough to be my son and old enough to be my brother, and he would have enjoyed the beer brewed with grains of paradise and lemon zest. I would have offered a chicken finger dipped in blue cheese, and he would have welcomed the crisp, tangy taste. We would have breathed the pine air and watched the shadows of the maples and white birch falling across the water as the sun floated down to the horizon. I would have placed a bite of filet mignon into his mouth, and he would have winked his approval. Later, I would have wheeled Matt inside the bathroom to drain his catheter bag of urine. Then, while the lake rippled away and the purple martins darted around in search of June bugs and the swans glided on the water, we would have sat on the deck for our last ale.

A year after Matt's kidney donation, I envisioned the nurse with lupus at the Cape, now confident that her body would not reject his dying gift. In the evening, her feet sandaled to prevent marine infections and jellyfish stings, she walks uphill on a sandy path, past the dried brush, past a dead tree that raises its brittle branches to the sky, past the goldenrod stems that sway in the evening breeze. Two tree swallows, their teal plumage merging with the bayberry, flutter around a bush as they share its starry fruit, the juice dripping down their beaks. They glide away in tandem. The nurse walks over the warm sand clinging to patches of beach grass, and at her left side she imagines a muscular young man with a smile flying across his face. She inhales the salt spray and sea lavender until the scent fills her lungs and enters the blood rushing through Matt's kidney, and when she exhales deeply, her breath rises into the dusk and floats away over the ocean.

She wrote a letter to the Nagles:

It was a very long journey, trying and at times extremely difficult. I waited for years for a kidney and during those years I often wondered if I would ever get that call, but I remained optimistic, and thanks to you I am able to live a normal life again. I am eternally grateful for your generosity and kindness, and I will never forget, ever.

The paralyzed boy in the electric wheelchair—whose thin legs had shaken in spasms during Matt's funeral service, whose arms moved just enough to control a knob on his chair—carries on the quest for his fallen comrade and continues racing with paralysis. I hope he keeps abreast of the latest research and best clinical practices for preventing urinary infections, kidney damage, pressure ulcers, bowel problems, osteoporosis, obesity, depression, and all the other complications of spinal cord injuries. I hope there is adequate funding for studies to prevent these complications, as well as for research that transforms paralysis into movement. I hope the boy goes on believing that exercise and sheer willpower, along with some miracle of science—whether brain implants or electric limbs or stem cells or even a healing combination—will raise up his arms and legs and he will roll away his electric chair and walk again, a baton grasped firmly in his hand as he wins the race for Matt.

Epilogue

S tem cell research continued after Matt's death. A 2007 congressio-
nal bill required the US Secretary of Health and Human Services to
conduct and support research with human embryonic stem cells under
stringent ethical guidelines. Embryos had to be donated by infertile
patients who no longer needed them, so they would never be implanted
and would certainly be destroyed. Donors had to give written, informed
consent and receive no financial or other inducements. Another section
required the secretary to assist with "the isolation, derivation, production, or testing of [alternative] stem cells that . . . may result in improved
understanding of treatments for diseases and other adverse health conditions."[1] True to his position in 2004, Barack Obama, then a senator,
strongly supported the bill. It passed in both the Senate and the House,
but President Bush vetoed it. He said that it was unethical to destroy
human life to help others, but he encouraged research with adult and
umbilical-cord stem cells.

Republican senator Arlen Specter had cosponsored the 2007 stem
cell legislation. On an earlier version, senator William Frist, a Republican physician, had commented, "I am pro-life, but I disagree with the

president's decision to veto the Stem Cell Research Enhancement Act. Given the potential of this research and the limitations of the existing lines eligible for federally funded research, I think additional [stem cell] lines should be made available."[2] Opponents of the bill said that alternative sources, such as umbilical cord and olfactory stem cells, should be used for research and treatment.

Martin Evans, who had been studying stem cells ever since Matt's birth, was a cowinner of the Nobel Prize for medicine in 2007. He had identified and isolated embryonic stem cells. He cultured them, genetically modified them, and reintroduced them into female rats to produce offspring that would breed an animal model for genetic defects. These achievements would allow scientists to understand the role of specific genes in human development, physiology, and pathology. And animal models are essential for research to develop treatments for human disease and disability.[3]

Geron Corporation announced that its human embryonic stem cells repaired spinal cord injuries in rats, without painful side effects, which validated "both the long-term safety and . . . effectiveness" of stem cells.[4] But in May 2008, the FDA rejected Geron's human clinical trial.[5] Its stock fell from five to two dollars over the next six months and then rose back to five after Barack Obama became president. Matt would have applauded the victory of this staunch supporter of stem cell research. With all the intensity of his football days, Matt would have rushed the political scrimmage line for Obama, blasting e-mails through his voice-activated computer and using his prominence as the first human Brain-Gate recipient to help elect a president who appeared to understand the suffering of millions of people with severe disabilities.

Geron submitted a revised protocol; it was twenty-one thousand pages long and had data from twenty-four animal studies. These animals showed no tumor formation, no neuropathic pain, and no increased mortality. Lab studies proved that the human immune system did not reject the stem cells. When the FDA approved the study in January 2009, Geron's stock rose from $5.50 to $8.50, but it came back down three months later, perhaps as investors realized that commercial stem cells could be far away and their clinical fate was uncertain.

President Obama's Executive Order 13505 in March 2009 stated that "research involving human embryonic and human non-embryonic stem cells has the potential to lead to better understanding and treatment of many disabling diseases and conditions." This research would be funded by the National Institutes of Health "if the cells were derived from human embryos that were created for reproductive purposes [and] were no longer needed for this purpose." Only embryos destined to be destroyed could be used for federally funded research, under strict guidelines and with written informed consent.[6]

"People are suffering," a priest told me. "If it were one of my loved ones, I'd have a much more difficult time responding to these questions about stem cell research. If it were my mother or my niece that really needed this research to walk or even to live, it would be very challenging. It's never just a simple clear-cut moral answer."

Scientists as well as priests hailed a study at the Scripps Research Institute: a special protein mixture induced the skin cells of adult mice to revert to an embryonic stage. In theory, this advance could limit or eliminate the need for embryonic stem cells.[7]

In August 2009, the FDA stopped Geron's first human trial, pending a review of data on cyst formation in rats treated with its embryonic stem cells. The company commented that "these cysts were non-proliferative, confined to the injury site, and had no adverse effects on the animals. No animals developed [tumors]. . . . Cysts of much larger size appear in the spinal cord scar tissue of up to 50 percent of patients with spinal cord injury."[8]

A 2010 research project at Scripps Research Institute showed that both embryonic and induced stem cells have genetic abnormalities. Embryonic stem cells have DNA duplication near genes that transform stem cells into other cells; induced ones have DNA deletions near genes for tumor suppression. These genetic defects are often related to cancers, so long-term monitoring is necessary if these cells are used to treat human beings.[9]

After the FDA was satisfied with animal data that addressed the concern about cysts, Geron resumed its trial in October 2010. The first person to receive an injection of about two million human embryonic stem cells for a spinal cord injury was at Shepherd Rehabilitation Center, where

Matt had been treated. The study's primary objective was to assess the safety of these cells in patients with complete paraplegia within fourteen days of spinal cord injuries. Matt would not have qualified for the clinical trial. In October 2011, Geron Corporation presented data on four patients with complete spinal cord injuries, including one who had been in the study for a year. There were no unexpected neurological changes, no cysts in the spinal cord MRI scans, and no immune responses to the stem cells. Mild side effects were due to an immunosuppressive drug used to prevent rejection of the stem cells.[10] The very next month, however, Geron announced that it would focus on cancer therapies and stop the spinal cord trial of stem cells. Geron would continue to monitor all enrolled patients and provide updates to the FDA. This shift was necessitated by "the current environment of capital scarcity and uncertain economic conditions."

In November 2010, an elderly man disabled by a stroke received a brain implant of neural stem cells. He was the first patient in ReNeuron's clinical trial at the Institute of Neurological Sciences in Glasgow.[11] Studies with rats had proved that the embryonic stem cells reduced inflammation, improved blood flow, and improved neurologic function after strokes. Twelve patients would receive a neurosurgical injection of those stem cells, with close monitoring for up to ten years to assess their safety and efficacy. Follow-up procedures would include blood tests to check for antibodies to the stem cells and MRI scans to check for bleeding, new strokes, inflammation, and tumors. In addition to neurological and cognitive examinations, the researchers would assess activities of daily living and health-related quality of life. In November 2011, ReNeuron reported that five patients had been treated with the stem cells and none had experienced any adverse events. It remains to be seen if the stem cells will help these patients recover from their strokes.

After umbilical stem cell therapy, Hwang Mi-Soon won a victory over paralysis and took a few steps with a walker. When the effects wore off, she had another treatment, but it led to an infection and chronic pain.

"I was like an animal they used for testing," she said in 2004.

Hwang's own doctor commented that the injection may have temporarily reduced her spinal pressure and that more animal research should have been done before anyone experimented with people. Back in her wheelchair, Hwang said she was speaking out so that others would not repeat her mistake.[12] (Matt had corresponded with Histostem, the company that had supplied Hwang's stem cells, but decided not to go to South Korea. People with paralysis are desperate—and vulnerable.)

What type of stem cells will be effective and safe for human patients? Currently, stem cells may be obtained from embryos, the umbilical cord, and adult tissues; genetic engineering through viruses or proteins may induce adult cells to act like stem cells. No one knows which source—if any—will help relieve human suffering. Only rigorous scientific experiments with long-term monitoring will give the answer. The contestants in this new scientific race have to proceed cautiously. From the medical and ethical perspectives, it would be ideal to treat patients with their own cells that have been transformed into an embryonic state: there would be no immune rejections and no use of embryos. These induced cells have to be carefully screened for genetic defects that might lead to cancer.

Research with the BrainGate also continued, with ambitious plans laid out on the company website: "Cyberkinetics is developing the Brain-Gate System to potentially provide limb movement to people with severe motor disabilities. In addition, Cyberkinetics is developing products to allow for robotic control, such as a thought-controlled wheelchair. Next generation products may be able to provide an individual with the ability to control devices that allow breathing, bladder and bowel movements." Investors contributed more than $40 million to this mission.

Almost two years after Matt's death, Cyberkinetics realized that it could not develop a commercial BrainGate device with its remaining funds, and so did investors. A *Boston Globe* article in March 2009 described the company as "winding down" its operations and "in the process of selling off its assets to pay creditors." Among the buyers was Jeff Stibel, a former neuroscience graduate student at Brown University and the president of the Internet company Web.com. He planned to form the BrainGate Company to further develop the software that decodes the messages of neurons.[13]

Although a fascinating achievement in neuroscience, the BrainGate was neither economically feasible nor clinically practical in its infancy. John Donoghue had listened to the music of brain neurons and heard the strings, the brass, and the woodwinds. He may have imagined the symphony that would make paralyzed muscles dance. Currently, with federal funding, he studies how to improve the BrainGate device, possibly with a wireless implant, in association with Spaulding Rehabilitation Hospital and Massachusetts General Hospital. Researcher Miguel Nicolelis also continues refining his knowledge of the nervous system and brain implants. The earliest cardiac pacemakers were as large as a toaster and sent impulses via thick wires. After decades of research funded by millions of dollars, modern devices have been miniaturized and made wireless; they are placed below the skin and remain effective for an average of seven years. Donoghue and Nicolelis continue their race toward a dream: if patients can use heart pacemakers, why not brain implants?

For now, people with severe disabilities use voice-activated and mouth-controlled devices to perform basic daily functions. And investors have moved on. The BrainGate won the half-marathon against stem cells in reaching the milestone of a human implant, but it was a costly victory.

Duke University and Miguel Nicolelis received a grant in October 2010 to develop a total-body prosthetic suit controlled by the brain's electrical waves. First, nerve signals would be recorded and interpreted; then these signals would be wirelessly transmitted to specific parts of the electronic suit, resulting in movement of the arms and legs. After monkey studies, Nicolelis planned a human trial to take place in 2014. Matt had successfully used the brain signals from his arm's motor cortex to control a virtual arm, but the Duke researchers set themselves the challenge of teaching a monkey—and eventually a human—to think about the *entire* body and make it move.

As with Matt's device, brain microelectrodes would record signals, but they would record the music of about a thousand neurons in the motor cortex. Eventually, Nicolelis and his team hope to record signals from fifty thousand individual nerves. An enormous amount of neural information is necessary for the complex process of moving the entire body, which requires the integration of speed, strength, and balance for harmonious muscle actions.

A paralyzed patient in the Duke study will learn to think about movements and make them happen on a computer screen, as Matt did. Once patients are able to control an avatar (a virtual person on a computer), they will wear an electronic suit that has sensors and muscle stimulators. In effect, the primate endoskeleton will be replaced by an exoskeleton. Brain signals will be sent wirelessly to sensors at strategic points in the prosthetic suit; then these signals will be used to stimulate coordinated movements.[14] In collaboration on a similar project, John Donoghue and Dr. Robert Kirsch (at the Cleveland FES Center) hope to someday coordinate a wireless BrainGate with an implanted electrode system. This might allow people with severe paralysis to use their brain signals for functional movement.[15] Matt would have qualified for this research in which brain implants are coordinated with functional electrical stimulation (FES).

Neurosurgeon Eric Leuthardt hopes to develop a brain-computer interface (BCI) like the BrainGate for patients with strokes. This is challenging, because brain damage due to the stroke makes nerve recordings unreliable or even impossible. A stroke in the left brain usually paralyzes the right side, but some nerves in the *right* brain are also involved in moving the limbs on the right. Distinct nerve patterns occur at the premotor cortex in the right frontal lobe, where movement is planned, before the actual signal to move the right hand. After a severe stroke on the left, a functional MRI scan shows increased activity in the premotor regions of the right hemisphere, as if the brain is struggling to plan movements. Leuthardt records these nerve signals with an electrode array between the brain surface and the scalp. (Because the brain's outer, fibrous covering is intact, there is a lower risk of brain infection than with an implant.) These signals could be used to control a robotic glove that opens and closes the hand for severely affected patients who have not benefited from intensive rehabilitation. In fact, Leuthardt's group at Washington University School of Medicine in Saint Louis has shown the feasibility of such a system.[16]

In 2010, the dreams of people with paralysis were renewed through major advances in stem cell research, neurological implants, and electric limbs. Matt would have been thirty-one years old. Jennifer French, whom I described in relation to Matt's desire for electrode implants controlled by the BrainGate, had computer-coordinated electrodes implanted in her leg and trunk muscles. In the summer of 2010, French had surgery for an advanced functional electrical stimulation system, with electrodes in more muscles, hoping for an improved pattern of standing. Two months later, the new system was turned on.

"My leg straightened out with full force," French told me in February 2011. "My therapist said I should become a football place kicker or a Rockette."

After the surgery, she embarked on an exercise program for muscle strength and endurance to prepare for standing.

"Body builders have nothing on me," she said with a wry smile.

Three months after her surgery, French resumed sailing and prepared to compete for the honor of representing America at the 2012 Paralympics in London. Then, at the end of 2010, after intensive physical exercises and complex technical adjustments, French stood up.

With the old system, her right knee had occasionally buckled.

"With the new system, I had a straight posture and solid knees," she commented. "And there was little use of my arms on the walker."

Implants near the spinal cord are the next step after muscle implants. Rob Summers, paralyzed below the waist after being struck by a car in 2006, received 150 hours of physical therapy while suspended in a harness, struggling to move his legs on a treadmill. There was no recovery. Then, in December 2009, researchers at the University of Louisville implanted a stimulator with sixteen electrodes over the spinal cord of the former college baseball player. The twenty-thousand-dollar device made by Medtronic is used for pain management, but the FDA approved it for spinal cord research in up to five patients.[17] In a series of sessions that lasted up to four hours, Summers was supported on a treadmill while the neuroscientists dialed up the current from one-half up to ten volts and tested frequencies from five to forty hertz. Simultaneously, the researchers recorded the electrical activity of fourteen lower limb muscles and the angles at the hip, knee, and ankle joints. This determined the optimal settings for electrical stimulation. With the device, Summers needed help to maintain his balance, but he stood up.

"I was able to stand independently the third day we turned it on,"[18] Summers said. "I was amazed."

For walking, the researchers programmed the spinal cord stimulator to generate rhythmic alternation of the lower limb muscles. Seven months after the surgery, Summers used the stimulator twice each day for voluntary control of some movements of his lower limbs. As described in the *Lancet* in May 2011, he had "robust, consistent rhythmic stepping-like activity" with assistance from a physical therapist.[19] This required continuous electrical stimulation and sensory feedback provided by weight bearing and bending and straightening of the legs. Summers's urinary function also improved, as reflected in the minimal amounts of urine left in the bladder after he voided. And the implant helped other problems associated with spinal cord injuries, including sexual function and temperature regulation. Summers said that his self-esteem, well-being, and social interactions had improved.

"I believe anything is possible," Summers said. "My ultimate goal is to walk and run again. I hope to someday go back to playing baseball."

Surgical implants are effective for restoring movement, but they have risks such as infection. To address this dilemma, Berkeley Bionics developed eLEGS, an exoskeleton strapped over clothing. The device, which includes a battery pack, weighs about forty-five pounds. It is similar to the ReWalk used by Floyd Morrow. Crutches stabilize the upright position, and patients need upper body strength to independently transfer to and from their wheelchairs. Matt would not have benefited from this technology. Sensors attached to the arms and legs determine the intention to move, and the exoskeleton's computer uses these data to move the legs forward, in spite of lower extremity paralysis. This system simulates a natural walking pattern, at a speed of about two miles per hour. A feat of mechanical engineering, electrical engineering, and programming, the exoskeleton is undergoing clinical research studies at select US rehabilitation programs. The director of the spinal cord injury program at the Reeve-Irvine Research Center—and an eLEGS user—Dr. Suzy Kim stated that the "eLEGS will revolutionize the field of neurologic rehabilitation." The eLEGS device was among *Time* magazine's fifty most important inventions of 2010, and CNN dubbed it the third-best innovation of the year.[20]

At the Cleveland FES Center, Dr. Ronald Triolo—inspired by Jennifer French and other former patients—is developing a hybrid system of implanted FES electrodes with exoskeleton braces to help patients with SCI to stand, walk, and even climb stairs. In concert with electrical stimulation of leg muscles, the device's joints lock (while the leg bears weight) and unlock (during the swing phase). The hip mechanism has better gait biomechanics than current braces but a slower speed, so Triolo plans to reduce the exoskeleton's weight. A hydraulic knee locks or unlocks based on foot pressure, knee angle, and muscle stimulation. An ankle joint moves the foot up or down, in synchrony with foot-to-floor contact and movement. Triolo hopes to improve knee-hip joint coordination, the fluidity of motion, and motorized assistance in this hybrid. In short, he wants the best of both FES and exoskeletal technologies.[21]

Noninvasive technologies keep improving, and some have recently won the da Vinci Award for assistive technology. One of the recipients was a hands-free speech-recognition system, called VoicePower Ultimate.[22] In the Tongue Drive system, a laptop wirelessly receives signals from a small magnetic device in the patient's tongue; for instance, touching an incisor with the tip of the tongue moves the wheelchair forward. The system also allows control of cell phones and home appliances.[23]

Electroencephalographic (EEG) recordings with scalp electrodes also hold promise for nonsurgical brain-computer interfaces. Specific EEG patterns are correlated with the patient's desire to control a computer cursor and used to select letters for word processing or icons for environmental control, as Matt did with his television. These recordings could also be converted into control signals for a brace or robotic device, like Matt's neuroprosthetic hand. Researchers including Dr. Jonathan Wolpaw at the Wadsworth Center in New York have proven that people with paralysis and absent sensation can modify their EEG wave patterns—as Matt did with his nerve signals—and use this mind control to drive a computer cursor. In a review article, Dr. Janis Daly of the Cleveland VA Medical Center and Wolpaw state that "the speed and precision of the multidimensional movement control achieved in human beings by sensorimotor-rhythm-based BCIs equals or exceeds that achieved so far with invasive methods."[24] The race between surgical and noninvasive technologies continues.

My specialty's journal, PM&R, devoted a special issue in December 2010 to neural plasticity. Guest editor Dr. Ross Zafonte of Spaulding Rehabilitation Hospital noted that "no single therapy will provide all the answers for our patients; rather, carefully adjusted combinations of therapy" would be more likely to offer the best outcomes.[25] His coeditor, Dr. David Chen of the Rehabilitation Institute of Chicago, expressed optimism about innovations that could "harness and use the capacity of the human nervous system to recover." But he was also concerned about "societal pressures" to study and apply "groundbreaking or controversial treatments, especially for conditions [with] a profound, life-changing impact . . . such as stroke, traumatic brain injury, spinal cord injury, and many other neurologic disorders." Chen believed that rehabilitation physicians had to "maintain scientific integrity and credibility" in order to "advocate for the best interest and outcomes for the patients we serve."[26]

In 2011, Bioness announced improvements in the devices used by Pat Wines and Kathy Spencer. The L300 Plus System has a thigh cuff for functional electrical stimulation of the hamstrings and quadriceps to help knee flexion and extension. And the H200 for the hand is now wireless, so the patient is no longer connected by wires to the control unit.[27]

I am glad to report that, as of spring 2012, Anna Iacono, Patricia Wines, Linda Holmes, Kathy Spencer, Garrett Mendez, Floyd Morrow, and Jennifer French were all thriving with their technologies in their race against paralysis.

Reflecting upon the experiences of these and other patients with spinal cord injury and stroke, I offer the following advice for adapting to, struggling against, and overcoming disabilities.

First, preserve all remaining nerve and muscle function. In the case of stroke, this means staying active to maintain muscles, avoiding drops or spikes in blood pressure, and staying on preventive medications such as blood thinners (if the stroke was not due to bleeding in the brain).

Second, prevent secondary complications such as blood clots (the leading cause of death in weeks two to four after a stroke), swallowing problems that can lead to pneumonia, and falls. For spinal cord injury, my greatest worries are urinary infections, which can lead to fatal septic shock, and pressure ulcers.

Third, take advantage of all traditional therapies, which are based upon almost a century of rehabilitation medicine. Physical, occupational, speech, and recreational therapists as well as rehabilitation nurses, case managers, and physicians can be invaluable coaches, allies, and advocates.

Fourth, try to create an optimal psychosocial environment. Familial and peer supports are essential to long-term survival. Disability organizations are invaluable resources, and they include the American Spinal Injury Association, the National Spinal Cord Injury Association, the American Stroke Association, the Brain Injury Association of America, the Multiple Sclerosis Society, the Muscular Dystrophy Association, and the American Parkinson Disease Association. The Michael J. Fox and the Christopher & Dana Reeve Foundations as well as other private foundations offer helpful advice. Disability studies programs are growing in number, and the educational institutions that offer such programs include Columbia University, Syracuse University, Temple University, and the University of Illinois at Chicago. Political advocacy for disability rights at the local and national levels is a crucial mission that involves many activists with disabilities.

Fifth, prevent long-term problems such as weight gain, which further compromises the ability to move with some degree of paralysis. A wellness program is essential. I discuss diet, exercise, smoking, alcohol, preventive medications, and lab tests with all my patients, and I send them off with a detailed letter.

Sixth, be open-minded *and* skeptical about rehabilitation equipment and technology. This book is a selective review, and there are other research projects as well as those in the planning stages—so choose carefully. When offered miracle cures, remember the stem cells of South Korea. Ask if the treatment will improve basic functions and quality of life. Inquire about the risks and alternative approaches. Start with noninvasive devices such as plastic braces and splints, Bioness devices, or the Myomo. Then proceed to minimally invasive technology, such as the implanted electrodes that Jennifer French used.

Finally, consider the most invasive devices only as a last resort. Explore these options to improve essential functions or to advance disability research—in the manner of Matt Nagle.

And, like Matt, never give up hope.

Acknowledgments

M any people offered their advice, support, and encouragement during the eight years in which I wrote this book. A research study like the BrainGate can succeed only with a brave person like Matthew Nagle, a dedicated research and clinical team, and numerous others. I am grateful to all who contributed.

First, I wish to thank Giselle Corré, who commented on the manuscript with the insight of a psychiatrist and the knowing eye of a wife. Her careful reading and perceptive comments have immeasurably improved this book. Giselle remains my favorite (literary) critic. For three decades, she has given me the gifts of support and understanding, but her greatest gifts to me are Nita and Jacob.

My children were remarkably tolerant as I worked on this book. Jacob offered helpful editorial advice, and Nita encouraged me to finish the book but did not unplug my computer, as she accidentally did at the age of two while I worked in the playroom on my PhD dissertation.

My mother, Ivy, was a teacher of English literature, and my father, Eric, was a physician. Their nurturing led to my love of literature and medicine, and I offer this book as a tribute to them.

Dwight Purdy, my first and perpetual English professor, reviewed at least three versions of this book. Ever since college, I have been grateful for his encouragement and critical guidance. On many occasions, he discussed everything from the title to the narrative structure of this book, offering wise counsel but letting me find my own way, as one would expect from a true teacher. I also benefited from Jeanne Purdy's skills as a teacher of writing and as a sounding board; revising this book was much easier with her warm support.

As a would-be writer, I married into the right family. My father-in-law, Alan Corré, an emeritus professor of linguistics, gave sage advice on the initial and final drafts. At a discouraging time for the publishing industry, Alan encouraged me to continue revising the book; he believed that it should and would be published.

My sister-in-law, Diana Newman, graciously commented on both the forest and the trees. A former editor and sophisticated reader, she helped improve this book with her long list of suggestions on the title, preface, and various narrative elements. I was also fortunate in having the legal and scholarly eye of my brother-in-law, Jacob Corré, who saved me from my inner scientific nerd.

It took a family to bring this book to its current state. Susan Beck, who is "family by choice," read drafts with the eye of a voracious reader and a sports fan. She kept me supplied with nonfiction books that were relevant to my work. I always paid careful attention to her insightful comments, which were often served with delicious dinners.

My good friend and talented artist, Tony Iorfino, has encouraged my writing ever since we were in medical school. I appreciated his enthusiasm for this project and his comments on one of the drafts.

Burke Barrett, my friend and colleague in the BrainGate study, played a pivotal role in the research project (and we had some good times while doing so). My readers and I are indebted to him for his careful review of and suggestions for improving an early draft.

I am grateful to John Donoghue and Tim Surgenor for inviting me to join the BrainGate project. John's discussions illuminated the complex technology that he developed as the father of the BrainGate, and I valued his historical perspective on neural prosthetics. Brown University is fortunate to have his leadership in the neurosciences.

Marilyn Serra, the president of Sargent Rehabilitation Center, graciously hosted the BrainGate study. She is a true believer in the mission of helping vulnerable people with disabilities.

At New England Sinai, I had illuminating discussions with Paula Picard, RN, and Brian Bloom, MD. At Rhode Island Hospital, the IRB was crucial in allowing the first BrainGate implant, and I am grateful to the board's members.

Leo Canuel commented on the book from his perspective as the executive director of a disability organization, PARI. Paul Choquette also helped me to better understand the viewpoints of people with disabilities.

Other scientists and clinicians provided valuable advice and thought-provoking discussions. They include the following people: Abe Caplan and Leigh Hochberg of Brown University; Hans Keirstead of the University of California, Irvine; Ronald Triolo, Kevin Kilgore, and Robert Kirsch of Cleveland FES Center; Shawn Wery of Cyberkinetics; Steve Williams of the University of Louisville; David Apple and Michael Jones of Shepherd Rehabilitation Center; Eric Leuthardt of the Washington University School of Medicine; and Joan Breen of Whittier Rehabilitation Hospital. My friend and fellow physiatrist, Jerrold Rosenberg, kindly provided coverage for my clinical work so that I could have time to write. I am indebted to my former resident Gary Polykoff, who generously served as my coinvestigator for the BrainGate study (and on short notice, too).

Gerhard Friehs helped me understand the neurosurgical details of the implant surgery. And he forever changed the meaning of *The Turn of the Screw*—at least in the annals of science—by tightening the nut (preamplifier) on the bolt (pedestal).

For religious perspectives on stem cells and disabilities, I am indebted to Bishop Thomas Tobin, Father Christopher Mahar, Reverend Gregory Stowe, and other members of the clergy. As a former professor of religious studies at Hofstra University, my cousin Vivodh Anand offered important advice on the ethical issues associated with stem cell research.

I was also fortunate to have the editorial help of Carol DeBoer-Langworthy of Brown University and her husband, Russell, and daughter, Sylvia Rolloff of Washington University, as well as Dorine Jennette and Leslie Rubinkowski. Each had a special perspective and gave detailed, critical advice.

In the humanities and medicine, I have been lucky to have the long-standing support and encouragement of Suzanne Poirier of the University of Illinois at Chicago and Lois LaCivita Nixon of the University of South Florida College of Medicine; both offered invaluable advice to improve this book. Other esteemed mentors in literature and medicine include Arthur Derse of Medical College of Wisconsin, Anne Hudson Jones of the University of Texas Medical Branch, Arnold Rampersad of Stanford University, and Howard Spiro of Yale University.

Some of Matt's friends were generous with their memories. Numerous phone discussions and visits with Katie Perette and Danny Taylor were essential to telling his story, and I am deeply grateful to both of them. I also enjoyed discussions with Jaryd Gray, Meaghan Murphy, Mike Romig, and Rich Rosado.

For their help with media and publicity, I owe thanks to Kari Watson of MacDougall Biomedical Communications; Pamela Getz of New England Sinai Hospital; Joanne Gemma, formerly of Sargent Rehabilitation Center; and Judi Hammerlind Carlson of TechACCESS of Rhode Island. This book would not be the same without color photographs and illustrations, and many people helped with this endeavor: Eilat Sinai-Cohen of Argo Medical Technologies, Jodi Hill of Bioness, Mary Buckett of Cleveland FES Center, Elizabeth Razee of Cyberkinetics, Bruno Dubuc of thebrain .mcgill.ca, Erin Perron of Monponsett Inn, Damien Woods of Moss-Rehab, Michael Quzor of Myomo, Barbara Murphy of National Academies Press, and Alfred Arcidi III of Whittier Rehabilitation Hospital.

Susan Betz welcomed me to Chicago Review Press, helped develop the ideas in this book, and offered wise editorial advice. I enjoyed many thought-provoking discussions with her. Kelly Wilson, the project editor for this book, went far beyond my expectations. Her insightful editorial comments and careful attention to myriad details—everything from checking facts to preparing the photographs and illustrations to organizing the text—have improved this book immensely. Natalya Balnova designed a beautiful and intriguing cover, and Sarah Olson's artistic interior design has also greatly enhanced the book. Marketing manager Mary Kravenas and publicist Meaghan Miller helped get the word out; I very much appreciate their enthusiasm, creativity, and energy. Cynthia Sherry, the publisher of Chicago Review Press, took on the challenge of this book, and I am thankful for her support.

I am deeply indebted to the patients in this book—and their families—who granted me the privilege of telling their stories and generously gave their time for long phone discussions and e-mail interchanges: Jennifer French, Linda Holmes, Anna Iacono, Garrett Mendez, Floyd Morrow, Kathy Spencer, and Patricia Wines. I admire these extraordinary people and wish them the best in their races against paralysis.

Ellen and Pat Nagle let me into their lives and spent many hours talking about their life with Matt as we looked over his memorabilia and photographs. I cannot thank them enough for their generosity and patience in discussing Matt's story. During these times, I enjoyed their company—and Ellen's meals. Their comments on the manuscript were invaluable.

Matt, you started and finished this book. I am so grateful for all you did for me and for people with disabilities. Thanks for the honest—and sometimes difficult—conversations. If you approve of this book, just wink.

Notes

Prologue

1. Hans Werner Pia, "Plasticity of the Central Nervous System: A Neuro-surgeon's Experience of Cerebral Compensation and Decompensation," *Acta Neurochirurgica* 77 (1985): 81–102.
2. Todd Neale, "ASA: Stroke Patients Getting Younger," MedPage Today, February 25, 2010, http://www.medpagetoday.com/MeetingCoverage /ASA/18677.
3. Segun T. Dawodu, "Spinal Cord Injury: Definition, Epidemiology, Pathophysiology," Medscape Reference, November 10, 2011, http:// emedicine.medscape.com/article/322480-overview.
4. Veronique L. Roger et al., "Heart Disease and Stroke Statistics—2011 Update: A Report from the American Heart Association," *Circulation* 123 (2011): 82–101.
5. "One Degree of Separation: Paralysis and Spinal Cord Injury in the United States," Christopher & Dana Reeve Foundation, accessed September 12, 2011, http://www.christopherreeve.org/site/c.mtKZKgMWKwG /b.5184255/k.6D74/Prevalence_of_Paralysis.htm.
6. Sue M. Lai et al., "Stroke Survival After Discharge from an Acute-Care Hospital," *Neuroepidemiology* 18 (1999): 210–17.

Chapter 2: The Hunting Knife at the Beach Party

1. "Spinal Cord Injury: A Brief History," in *Spinal Network: The Total Wheelchair Resource Book*, ed. Barry Corbett, Jean Dobbs, and Bob Bonin (Santa Monica, CA: Nine Lives Press, 2002), 59–64.

Chapter 3: Stairway to Recovery

1. Julius P. A. Dewald, "Sensorimotor Neurophysiology and the Basis of Neurofacilitation Therapeutic Techniques," in *Stroke Rehabilitation*, ed. Murray E. Brandstater and John V. Basmajian (Baltimore, MD: Williams and Wilkins, 1987), 109–82.

Chapter 4: Alive in the Electric Chair

1. "Spinal Cord Injury: A Brief History," in *Spinal Network: The Total Wheelchair Resource Book*, ed. Barry Corbett, Jean Dobbs, and Bob Bonin (Santa Monica, CA: Nine Lives Press, 2002), 59–64.

Chapter 5: Parade with the Patriots

1. "Human Cloning," by Susan Dentzer, *NewsHour with Jim Lehrer*, PBS, November 26, 2001, http://www.pbs.org/newshour/bb/health/july-dec01/cloning_11-26.html.

Chapter 6: Riding with an Electric Leg

1. Jeffrey M. Hausdorff and Haim Ring, "Effects of a New Radiofrequency–Controlled Neuroprosthesis on Gait Symmetry and Rhythmicity in Patients with Chronic Hemiparesis," *American Journal of Physical Medicine and Rehabilitation* 87 (2008): 4–13.
2. Yocheved Laufer, Jeffrey M. Hausdorff, and Haim Ring, "Effects of a Foot Drop Neuroprosthesis on Functional Abilities, Social Participation, and Gait Velocity," *American Journal of Physical Medicine and Rehabilitation* 88 (2009): 14–20.
3. Shaheen Hamdy et al., "Long-Term Reorganization of Human Motor Cortex Driven by Short-Term Sensory Stimulation," *Nature Neuroscience* 1 (1998): 64–68.
4. Alain Kaelin-Lang et al., "Modulation of Human Corticomotor Excitability by Somatosensory Input," *Journal of Physiology* 540 (2002): 623–33.
5. Joachim Liepert et al., "Treatment-Induced Cortical Reorganization After Stroke in Humans," *Stroke* 31 (2000): 1210–16.

Chapter 8: Monkeys Playing Video Games

1. E. Maynard et al., "Simultaneous Recordings of Motor Cortical Neurons Allow Estimation of Movement Direction from Small Numbers of Neurons," abstract (presented at the Society for Neuroscience annual meeting, Washington, DC, November 1996).
2. Liam Paninski et al., "Coding Dynamic Variables in Populations of Motor Cortex Neurons," abstract (presented at the Society for Neuroscience annual meeting, Miami Beach, FL, October 1999).
3. Mijail Serruya et al., "Instant Neural Control of a Movement Signal," *Nature* 416 (2002): 141–42.
4. Cyberkinetics reviewed relevant parts of this book.
5. "Selected FDA GCP/Clinical Trial Guidance Documents," U.S. Food and Drug Administration, last modified March 5, 2012, http://www.fda.gov/ScienceResearch/SpecialTopics/RunningClinicalTrials/GuidancesInformationSheetsandNotices/ucm219433.htm.

Chapter 9: A Walk Around the Lake

1. "Keeping Your Balance," excerpted and adapted from "Keeping Your Balance, Everyday Survival," *Stroke Connection Magazine*, July/August 2003, last modified August 11, 2010, http://www.strokeassociation.org/STROKEORG/LifeAfterStroke/RegainingIndependence/PhysicalChallenges/Keeping-Your-Balance_UCM_309772_Article.jsp#.T1pklHlx1-M.
2. "Patient BP Case History," Tibion, accessed June 2011, http://www.soundwebdesign.net/tibtest/?page_id=167.
3. John Burn et al., "Long-Term Risk of Recurrent Stroke After a First-Ever Stroke: The Oxfordshire Community Stroke Project," *Stroke* 25 (1994): 333–37.

Chapter 10: Navigating a Study Through Two Research Review Boards

1. "Testimony of Vladislava Karolewska," from "The Nuremberg Trials: The Doctor Trials: Transcript Excerpts," University of Missouri–Kansas City School of Law, accessed July 2011, http://law2.umkc.edu/faculty/projects/ftrials/nuremberg/NurembergDoctorTranscript.html.
2. "U.S. Public Health Service Syphilis Study at Tuskegee: The Tuskegee Timeline," Centers for Disease Control and Prevention, last modified June 15, 2011, http://www.cdc.gov/tuskegee/timeline.htm.
3. The National Commission for the Protection of Human Subjects of Biomedical and Behavioral Research, *The Belmont Report: Ethical Principles and Guidelines for the Protection of Human Subjects of Research* (Bethesda,

MD: National Institutes of Health Office of Human Subjects Research, 1979), http://ohsr.od.nih.gov/guidelines/belmont.html.

4. Melanie Fried-Oken, "Voice Recognition Device as a Computer Interface for Motor and Speech Impaired People," *Archives of Physical Medicine and Rehabilitation* 10 (1985): 678–81.

5. Andrew Pollack, "With Tiny Brain Implants, Just Thinking May Make It So," *New York Times*, April 13, 2004.

6. Felice Freyer, "Testing the Power of Thoughts," *Providence Journal*, April 21, 2004.

Chapter 11: Play It Again

1. "Regain More Natural Hand Function," Bioness, accessed July 2011, http://www.bioness.com/H200_for_Hand_Paralysis.php.

2. Gad Alon et al., "A Home-Based, Self-Administered Stimulation Program to Improve Selected Hand Functions of Chronic Stroke," *Neuro-Rehabilitation* 18 (2003): 215–25.

3. Thomas E. Twitchell, "The Restoration of Motor Function Following Hemiplegia in Man," *Brain* 74 (1951): 443–80.

4. Paul Bach-y-Rita, "Process of Recovery from Stroke," in *Stroke Rehabilitation*, ed. Murray E. Brandstater and John V. Basmajian (Baltimore, MD: Williams and Wilkins, 1987), 80–108.

5. Albrecht T. Julius Bethe, *Allgemeine Anatomie und Physiologie des Nervensystems* (Leipzig, Germany: Thieme, 1910).

6. Paul Weiss and Paul F. Brown, "Electromyographic Studies on Recoordination of Leg Movements in Poliomyelitis Patients with Transposed Tendons," *Proceedings of the Society of Experimental Biology* 48 (1941): 284–87.

Chapter 13: Wired, Connected, and Waiting for Signals

1. Jon Mukand et al., "Feasibility Study of a Neural Interface System for Quadriplegic Patients," *Archives of Physical Medicine and Rehabilitation* 85 (2004): 48.

2. Masayoshi Ohta et al., "Bone Marrow Stromal Cells Infused into the Cerebrospinal Fluid Promote Functional Recovery of the Injured Rat Spinal Cord with Reduced Cavity Formation," *Experimental Neurology* 187 (2004): 266–78.

3. Daniel P. Ankeny, Dana M. McTigue, and Lyn B. Jakeman, "Bone Marrow Transplants Provide Tissue Protection and Directional Guidance for Axons After Contusive Spinal Cord Injury in Rats," *Experimental Neurology* 190 (2004): 17–31.

4. Steven Ertelt and Maria Vitale Gallagher, "Paralyzed South Korean Woman Walks Thanks to Adult Stem Cell Research," LifeNews.com, November 29, 2004, http://www.lifenews.com/2004/11/29/bio-582/.
5. "Real Stories," Neurotech Network, accessed July 2011, http://www.neurotechnetwork.org/education/real_stories.htm.
6. Ibid.
7. Robert Kirsch, PhD, and Ronald Triolo, PhD, phone interview by the author, November 22, 2011.

Chapter 14: Skating Away from a Stroke

1. "Brain Stem Stroke," excerpted and adapted from "Surviving a Brain Stem Stroke," *Stroke Connection Magazine*, January/February 2003, last modified March 6, 2012, http://www.strokeassociation.org/STROKEORG/AboutStroke/EffectsofStroke/Brain-Stem-Stroke_UCM_310771_Article.jsp.
2. Jean-Dominique Bauby, *The Diving Bell and the Butterfly: A Memoir of Life in Death* (New York: Alfred A. Knopf, 1997), 32.
3. Kate Laver et al., "Virtual Reality for Stroke Rehabilitation," *Cochrane Database of Systematic Reviews* 9 (2011), doi: 10.1002/14651858.CD008349.pub2.
4. "Walk Aide," YouTube video, 4:23, posted by "entourage1000," January 2, 2010, http://www.youtube.com/watch?v=nsIBxUwxFIg.
5. "What Is the *Saebo*Flex?," Saebo, accessed May 2011, http://www.saebo.com/products/saeboflex/.
6. Jeffery Kurz, "Stroke of Genius," *Record-Journal* (Meriden, CT), May 17, 2008, http://www.myrecordjournal.com/latestnews/article_f34c49d0-ba60-591c-928d-60c8023bdcb2.html.
7. "Myomo Mobility System: mPower 1000," Myomo, accessed May 2011, http://myomo.com/myomo-solutions-mpower-1000/.
8. Ibid.
9. Ibid.
10. Steve Heath, "Hockey Aids Recovery," USAHockey.com, May 19, 2008, http://hockey.teamusa.org/news/2008/05/22/hockey-aids-recovery/63.
11. Garrett Mendez ("entourage1000") YouTube page, accessed May 2011, http://www.youtube.com/results?search_query=entourage1000&aq=0.
12. "Skating December 22, 2009," YouTube video, 2:04, posted by "entourage1000," December 24, 2009, http://www.youtube.com/watch?v=u1z9yiQGhSU.

Chapter 15: The Bionic Man

1. Leigh R. Hochberg et al., "Neuronal Ensemble Control of Prosthetic Devices by a Human with Tetraplegia," *Nature* 442 (2006): 164–71.

Chapter 16: Imagining a God

1. Leander Kahney, "Biggest Discoveries of 2005," *Wired*, December 29, 2005, http://www.wired.com/science/discoveries/news/2005/12/69909 ?currentPage=all.
2. Jeannette E. Davies et al., "Astrocytes Derived from Glial-Restricted Precursors Promote Spinal Cord Repair," *Journal of Biology* 5 (2006): 7.
3. Gareth Cook, "Chinese Surgeon's Claims about Cell Implants Disputed," *Boston Globe*, June 19, 2006.
4. Bruce H. Dobkin, A. Curt, and J. Guest, "Cellular Transplants in China: Observational Study from the Largest Human Experiment in Chronic Spinal Cord Injury," *NeuroRehabilitation and Neural Repair* 20 (2006): 5–13.
5. Sheryl Gay Stolberg, "Nancy Reagan Supports Stem Cell Bill," *New York Times*, May 16, 2006, http://www.nytimes.com/2006/05/16 /washington/16brfs-brief-006.html?_r=1.
6. Andrea Stone, "Limbaugh Says Actor Fox Exaggerating His Disease as Stem Cell Issue Churns," *USA Today*, October 25, 2006, http://www.usa today.com/news/politicselections/2006-10-24-limbaugh-fox_x.htm.
7. Jeffrey V. Rosenfeld and Grant R. Gillett, "Ethics, Stem Cells and Spinal Cord Repair," *Medical Journal of Australia* 180 (2004): 637–39.

Chapter 17: A Starship Trooper's Electronic Suit

1. "Science Behind Wound Therapy," KCI, accessed July 2011, http:// www.kci1.com/KCI1/sciencebehindthetherapy#howworks.
2. Maurizio Massucci et al., "Walking with the Advanced Reciprocating Gait Orthosis (ARGO) in Thoracic Paraplegic Patients: Energy Expenditure and Cardiorespiratory Performance," *Spinal Cord* 36 (1998): 223–27.
3. "*Starship Troopers*," Wikipedia, last modified March 13, 2012, http:// en.wikipedia.org/wiki/Starship_Troopers.
4. They Shall Walk website, accessed March 2011, http://theyshallwalk .org.
5. "Technology," Argo Medical Technologies, accessed March 2011, http:// www.argomedtec.com/technology.asp.
6. Suzan Clarke, "Cutting-Edge Robotic Exoskeleton Allows Wheelchair-Bound to Stand and Walk," ABC News, February 4, 2010, http://

abcnews.go.com/GMA/OnCall/bionic-breakthrough-robotic-suit
-helps-paraplegics-walk/story?id=9741496&page=2.

7. Katie Taylor, "Glee Trousers: The Next Step," *Jewish Chronicle*,
December 16, 2010, http://www.thejc.com/news/uk-news/42699
/glee-trousers-next-step.

Chapter 18: In Flight

1. Brown University, "Brown Scientist John P. Donoghue Wins Major
Neuroscience Award," news release, August 20, 2007, http://news
.brown.edu/pressreleases/2007/08/neuroscience-award.

2. Cure Paralysis Now forum, accessed December 2010, http://www.cure
paralysisnow.org/forum/viewtopic.php?t=1143.

3. Seung Hwan Yoon et al., "Complete Spinal Cord Injury Treatment
Using Autologous Bone Marrow Cell Transplantation and Bone Mar-
row Stimulation with Granulocyte Macrophage-Colony Stimulating
Factor: Phase I/II Clinical Trial," *Stem Cells* 25 (2007): 2066–73.

4. Monya De, "New Stem Cell Breakthrough Avoids Destroying Human
Embryos," ABC News, June 6, 2007, http://abcnews.go.com/Health
/Technology/story?id=3250281&page=1.

5. Roger Highfield, "Stem Cell Research Revolution Spells End for Therapeutic Cloning," *Telegraph* (London), November 20, 2007, http://www
.telegraph.co.uk/science/science-news/3315273/Stem-cell-research
-revolution-spells-end-for-therapeutic-cloning.html.

6. Monya De, "New Stem Cell Breakthrough."

7. Rick Weiss, "Advance May End Stem Cell Debate," *Washington Post*,
November 21, 2007, http://www.washingtonpost.com/wp-dyn/content
/article/2007/11/20/AR2007112000546.html.

8. Jeff Biernaskie et al., "Skin-Derived Precursors Generate Myelinating
Schwann Cells That Promote Remyelination and Functional Recovery
After Contusion Spinal Cord Injury," *Journal of Neuroscience* 5 (2007):
9545–59, http://www.jneurosci.org/content/27/36/9545.abstract.

Chapter 19: Aflame in Red Chrysanthemums

1. The material in italics is excerpted from *Celebrating the Eucharist* (2007)
and inspired by hymns that are especially relevant to people with disabilities. These hymns include Daniel L. Schutte, "Here I Am, Lord,"
and Davis Haas, "You Are Mine," from Oregon Catholic Press, *Breaking
Bread 2012 Missal* (Portland, OR: Oregon Catholic Press, 2012), pp. 375,
491.

Epilogue

1. Stem Cell Research Enhancement Act of 2007, S. 997, 110th Cong. (2007). http://www.opencongress.org/bill/110-s997/text.

2. Joanne Carney, "Senate Passes Stem Cell Bill," American Association for the Advancement of Science newsletter (July 2006), http://www .aaas.org/gr/stc/Archive/stc06/06_07_stcnewsletter.shtml#Stemcells.

3. "The Nobel Prize in Physiology or Medicine 2007," Nobel Prize official website, last modified March 14, 2012, http://www.nobelprize.org /nobel_prizes/medicine/laureates/2007/press.html.

4. Geron, "Data Show Geron's Cell-Based Therapeutic for Spinal Cord Injury Survives and Exhibits Remyelination for at Least Nine Months Following Injection," news release, November 7, 2007, http://ir.geron .com/phoenix.zhtml?c=67323&p=irol-newsArticle_Print&ID=1636828 &highlight=.

5. Geron, "FDA Places Geron's GRNOPC1 IND on Clinical Hold," news release, May 14, 2008, http://ir.geron.com/phoenix.zhtml?c= 67323&p=irol-newsArticle&ID=1636306&highlight=.

6. Exec. Order 13505, 74 Fed. Reg. 10667 (March 11, 2009).

7. Bradley J. Fikes, "Medicine: Scripps Scientists Create Safer Artificial Stem Cells," *North County Times* (San Diego, CA), April 27, 2009, http://www .nctimes.com/news/science/medicine-scripps-scientists-create-safer-arti ficial-stem-cells/article_641e3a21-d5b1-5dd7-91d4-48b424a9509c.html.

8. Geron, "Geron Comments on FDA Hold on Spinal Cord Injury Trial," news release, August 27, 2009, http://ir.geron.com/phoenix .zhtml?c=67323&p=irol-newsArticle&ID=1636251&highlight=.

9. "Genetic Abnormalities Identified in Pluripotent Stem Cell Lines," PhysOrg.com, January 6, 2011, http://www.physorg.com/news/2011 -01-genetic-abnormalities-pluripotent-stem-cell.html.

10. Geron, "Geron Presents Clinical Data Update from GRNOPC1 Spinal Cord Injury Trial," news release, October 20, 2011, http://ir.geron.com /phoenix.zhtml?c=67323&p=irol-newsArticle&ID=1635760&highlight=.

11. ReNeuron, "ReNeuron Announces First Patient Treated in Landmark Stroke Stem Cell Clinical Trial," news release, November 11, 2010, http://www.reneuron.com/press-release/reneuron-announces-first -patient-treated-in-landmark-stroke-stem-cell-clinical-trial-16-11-10.

12. Barbara Demick, "Faith in 'Miracle Cures' Is Fading in South Korea," *Los Angeles Times*, March 5, 2006, http://articles.latimes.com/2006 /mar/05/world/fg-stemcell5.

13. Scott Kirsner, "CyberKinetics' Brain-to-Computer Interface Gets a Second Chance," *Boston Globe*, August 12, 2009, http://www.boston.com

/business/technology/innoeco/2009/08/cyberkinetics_braintocom
puter.html.

14. Shirley S. Wang, "Wiring the Brain to Aid People with Paralysis," *Wall Street Journal*, October 26, 2010, http://online.wsj.com/article/SB100014240
52702304248704575574181134954298.html.

15. Cleveland FES Center, "Robert Kirsch, PhD, Elected to AIMBE College of Fellows," news release, accessed November 2009, http://fescenter
.org/index.php?view=article&id=97%3Arobert-kirsch-phd-elected-to
-aimbe-college-of-fellows&option=com_content&Itemid=19.

16. Anthony Ritaccio et al., "Proceedings of the First International Workshop on Advances in Electrocorticography," *Epilepsy & Behavior* 19 (2010): 204–15.

17. Susan Harkema et al., "Effect of Epidural Stimulation of the Lumbosacral Spinal Cord on Voluntary Movement, Standing, and Assisted Stepping After Motor Complete Paraplegia: A Case Study," *Lancet* 377 (2011): 1938–47.

18. Silvia Pikal, "Paralyzed Man Stands and Takes Steps After Spinal Implant," *Mobile*, May 21, 2011, http://www.mobilemag.com/2011/05
/21/paralyzed-man-stands-and-takes-steps-after-spinal-implant/.

19. Susan Harkema, "Effect of Epidural Stimulation of the Lumbosacral Spinal Cord on Voluntary Movement, Standing, and Assisted Stepping After Motor Complete Paraplegia: A Case Study," *Lancet* 377 (2011): doi:10.1016
/S0140-6736(11)60547-3.

20. Alice Park, "The 50 Best Inventions of 2010: eLegs Exoskeleton," *Time*, November 11, 2010, http://www.time.com/time/specials/packages/article
/0,28804,2029497_2030618_2029794,00.html.

21. *VA Research Today*, "APT Center Is Home to Emerging Technology for Veterans with Disabilities," Spring 2010, http://www.research.va.gov
/news/features/VA-ResearchToday.pdf, 2–5.

22. "2010 Winners," da Vinci Awards website, accessed December 2010, http://
www.davinciawards.org/about/winners/?year=2010.

23. Georgia Institute of Technology, "Tongue Drive System Selected as 2010 da Vinci Awards Finalist," news release, September 13, 2010, http://www
.gatech.edu/newsroom/release.html?nid=60941.

24. Janis J. Daly and Jonathan R. Wolpaw, "Brain–Computer Interfaces in Neurological Rehabilitation," *Lancet* 7 (2008): 1032–43.

25. Ross Zafonte, "Introduction," *PM&R: The Journal of Injury, Function and Rehabilitation* 2, supplement 2 (2010): s207.

26. David Chen, "Neural Plasticity: Future Controversies?," *PM&R: The Journal of Injury, Function and Rehabilitation* 2, supplement 2 (2010): s313–4.

27. "How the NESS H200 Hand Rehabilitation System Works," Bioness, accessed December 2011, http://www.bioness.com/United_Kingdom
/H200_Wireless_for_Hand_Paralysis/How_Does_It_Work.php.

Index